Leonardo's Mirror
& other Puzzles

IVAN MOSCOVICH

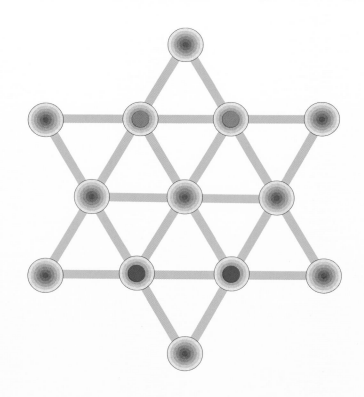

Dover Publications, Inc.
Mineola, New York

To Anitta, Hila, and Emilia, with love

Bibliographical Note

This Dover edition, first published in 2011, is an unabridged republication of the work originally published in 2004 by Sterling Publishing Company, Inc., New York.

Library of Congress Cataloging-in-Publication Data

Moscovich, Ivan.
    Leonardo's mirror and other puzzles / Ivan Moscovich
        p. cm.
    Originally published: New York : Sterling Publishing Co., 2004
    ISBN-13: 978-0-486-48239-2
    ISBN-10: 0-486-48239-1
    1. Puzzles 2. Scientific recreations. I. Title
GVI493.M6125 2011
793.73--dc22

2010054013

Manufactured in the United States by Courier Corporation
48239101
www.doverpublications.com

Contents

Introduction

Ever since my high-school days I have loved puzzles and mathematical recreational problems. This love developed into a hobby when, by chance, some time in 1956, I encountered the first issue of *Scientific American* with Martin Gardner's mathematical games column. And for the past 50 years or so I have been designing and inventing teaching aids, puzzles, games, toys and hands-on science museum exhibits.

Recreational mathematics is mathematics with the emphasis on fun, but, of course, this definition is far too general. The popular fun and pedagogic aspects of recreational mathematics overlap considerably, and there is no clear boundary between recreational and "serious" mathematics. You don't have to be a mathematician to enjoy mathematics. It is just another language, the language of creative thinking and problem-solving, which will enrich your life, like it did and still does mine.

Many people seem convinced that it is possible to get along quite nicely without any mathematical knowledge. This is not so: Mathematics is the basis of all knowledge and the bearer of all high culture. It is never too late to start enjoying and learning the basics of math, which will furnish our all-too sluggish brains with solid mental exercise and provide us with a variety of pleasures to which we may be entirely unaccustomed.

In collecting and creating puzzles, I favour those that are more than just fun, preferring instead puzzles that offer opportunities for intellectual satisfaction and learning experiences, as well as provoking curiosity and creative thinking. To stress these criteria, I call my puzzles Thinkthings.

The *Mastermind Collection* series systematically covers a wide range of mathematical ideas, through a great variety of puzzles, games, problems, and much more, from the best classical puzzles taken from the history of mathematics to many entirely original ideas.

This book, *Leonardo's Mirror & Other Puzzles*, contains a tribute to Leonardo da Vinci, undoubtedly the most creative person our world has known. As well his famous hidden message, it contains several of his dissection problems, found amid a vast array of other popular and novel puzzles, games, and more.

A great effort has been made to make all the puzzles understandable to everybody, though some of the solutions may be hard work. For this reason, the ideas are presented in a novel and highly esthetic visual form, making it easier to perceive the underlying mathematics.

More than ever before, I hope that these books will convey my enthusiasm for and fascination with mathematics and share these with the reader. They combine fun and entertainment with intellectual challenges, through which a great number of ideas, basic concepts common to art, science, and everyday life, can be enjoyed and understood.

Some of the games included are designed so that they can easily be made and played. The structure of many is such that they will excite the mind, suggest new ideas and insights, and pave the way for new modes of thought and creative expression.

Despite the diversity of topics, there is an underlying continuity in the topics included. Each individual Thinkthing can stand alone (even if it is, in fact, related to many others), so you can dip in at will without the frustration of cross-referencing.

I hope you will enjoy the *Mastermind Collection* series and Thinkthings as much as I have enjoyed creating them for you.

—Ivan Moscovich

Endless variations of the basic curves—forms that are not quite circles, ellipses, or parabolas, etc.—are around us everywhere. While we associate violence with angles and broken, jagged lines, we equate restfulness with horizontal lines and gradual curves, and motion with curves that continuously change direction.

✳ Curves in nature

A curve is a line that continuously bends but has no angles. Some, such as parabolas, are open: That is, the line never returns to its starting point. Some join up with themselves; an example is an oval, which is closed.

Some curves are twisted, such as the helix. A heavy chain left to hang between two points forms a natural curve known as the catenary (see page 56). The path made by a point on a wheel as it rolls along a flat surface is called a cycloid. The shape of an airplane wing and the paths of rockets are special curves determined by mathematics.

Some curves take the shortest path possible. The problem of finding a shape of "minimal surface" or "minimal curve" within a given boundary is known as Plateau's problem. In over 180 years of research, it is still mathematically unsolved.

Soap bubbles are examples of minimal surfaces, so called because nature selects the shape that requires the least amount of energy to enclose a given area or volume with as little surface or perimeter as possible.

Have you seen a straight river? Most unlikely. In fact, the repeating pattern of turns is, in most instances, the most distinctive feature of a river—such curves are called meanders. Snakes, rivers, and many other natural phenomena tend to move in a wavy, meandering line. The almost geometric regularity of river meanders has attracted the interest of scientists for many years. It is no accident: Meanders appear to be the form in which a river does the least work in the process of turning.

A strip of thin steel can be bent into various configurations—all of which are models of river meanders. When holding the strip firmly at two points, it assumes the shape in which the bend is as uniform as possible. And what is that bend? A curve—a line that continuously bends but has no angles.

ANALEMMA
Curve in the sky

This famous time-lapse photo by Dennis di Cicco is the result of 44 exposures taken every seven or eight days or so at the same time of the day, over a period of exactly one year. It shows the sun tracing a figure eight curve in the sky.

Can you explain the reasons for the strange curve? Why are its loops unequal?

In a perfect world in which the Earth would orbit the sun in a perfect circle, with its equator on the same plane as its orbit, how would this photo look?

(D. di Cicco, Sky Publishing Corporation, 1979)

(ANSWER: PAGE 98)

Concentric stars

You may have seen photographs of the night sky like this one on the right.

Can you explain the concentric paths of the stars and work out how the photograph was produced?

(ANSWER: PAGE 98)

If the Lord Almighty had consulted me before embarking upon the Creation, I should have recommended something simpler.
Alfonso X, El Sabio (The Wise) of Castile (1221–1284), upon being instructed in the Ptolemaic system

Complex planetary curves

By the 4th century B.C. heavenly bodies were already being studied. In the famous Library of Alexandria, volumes were accumulated about the motion of heavenly bodies. Records showed that heavenly bodies moved in two categories.

The sun, the moon, and the contained sphere of stars orbited the Earth in near-circular paths. But the five (then) known planets—Saturn, Jupiter, Mars, Venus, and Mercury—moved in mysterious looped paths. One night a planet can be seen moving in one direction, and, later, can be seen in another. But why was this so?

Their paths are in fact epicyclic as shown (epicyclic curves occur when a circle rolls around inside another, a little like watching one point of a coin as it revolves inside a larger ring). The photo on the right taken in a planetarium shows the complex paths of the five planets. Can you explain why we see the paths of outer planets as looped epicyclic curves from the Earth?

(ANSWER: PAGE 99)

You may have seen threaded designs like these before on pinboards. They're pretty easy to construct on paper, with a little patience.

▶ STRING ALONG 1

As shown by the small diagram on the right, start with a chord (line) across the circle then move one end by one notch each time, and the other end by two notches. This 1:2 ratio generates a cardioid (heart) curve, as shown.

Patterns and curves from straight lines

Another way to produce these patterns is to copy and cut out the templates from stiff cardboard and slot them at equally spaced intervals. Then, use colored string to connect the points along the slots, producing striking patterns. This is an ancient craft technique called "curve stitching," which is educationally meaningful because of its mathematical relevance. Series of straight lines produce the illusions of striking curves that are touched by the straight lines. In mathematical language, the curves are "envelopes" of the straight lines. This mathematical property makes these designs simple to produce and rewarding in their surprising results.

You can easily invent your own rules, templates, and experiments. (To save time, you can use circles with fewer dots around the circumference if you like—for example, 36 dots at 10 degrees spacing—but your curves will be a little less smooth.)

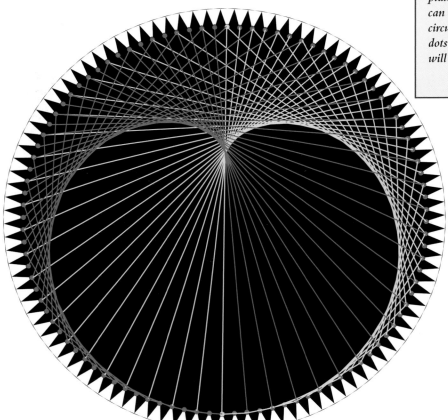

▶ STRING ALONG 2

As shown on the right, again start with a line across the circle, but this time move one side one notch per turn and move the other end by three notches. What curve does this 1:3 rule generate?

ANSWER: PAGE 99

If you've ever seen puzzles that ask you things like "What is the largest number of areas you can form using three straight lines?," the techniques on this page will teach you the best way of solving them. So, forewarned is forearmed!

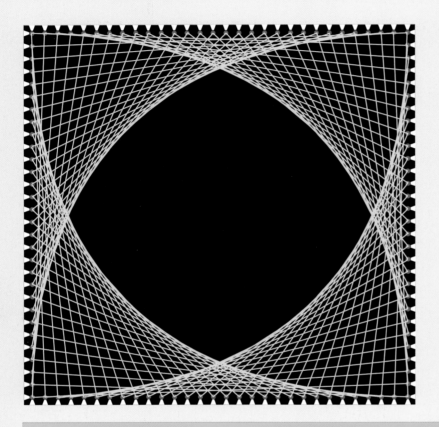

Rule I

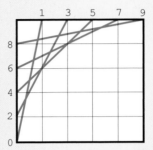

Note how these five lines divide the square into 16 areas—the most possible.

✳ Line designs

On these pages you will find examples of patterns and curves created from straight lines through curve-stitching (see also page 8).

Curve-stitching is based on geometrical forms, creating the illusion of curves and circles from straight lines. It is a rewarding craft activity, combining geometry and number patterns with a strong emphasis on order and symmetry. The example above shows the pattern produced by following the technique illustrated beside it. Here the sides of a square frame are divided into equally spaced notches. In step 1, point 0 is con-nected to the first notch of the adjacent side go-ing clockwise. In step 2, point 2 is connected to the second point on the adjacent side, and so on. This is the 1:1 rule.

Now try the example on page 11 yourself. Rule 2 follows the same method as rule 1 (1,1) but starts in the middle of a side instead of the corner, while in rule 3, consecutive points on one side are connected to every second point on the adjacent side (1:2). But the real fun lies in devising your own rules and seeing the outcomes.

▼ STRING ALONG 3

This grid will enable you to produce a startling pattern. Choose rule 2 or 3 from below and apply it four times, once in each corner of the grid. Can you envisage what it will look like before you begin?

ANSWER: PAGE 99

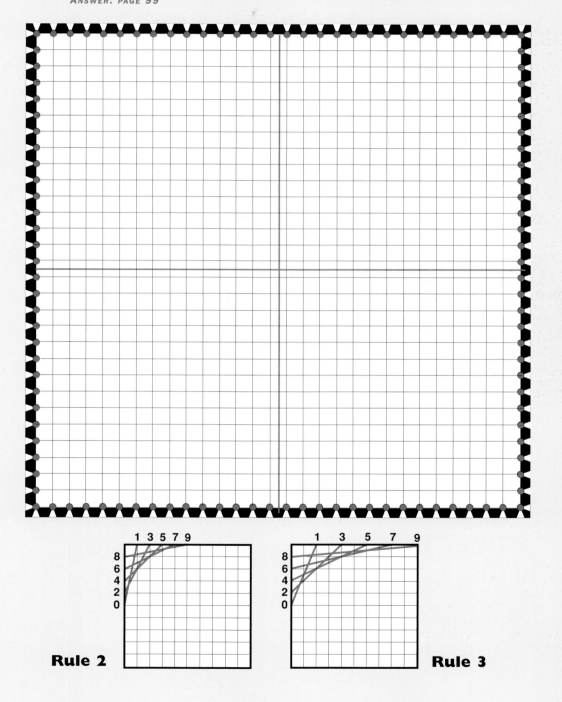

Rule 2

Rule 3

The Greek tradition of smashing plates in restaurants is associated with the *Kefi*, a friendly celebration of food, wine, and fun. Or maybe they just really enjoy jigsaws. You'll have your work cut out for you with these teasers.

⊠ BROKEN NECKLACE

The precious necklace above was broken into nine pieces.

Copy and cut out the pieces, then reassemble them into a closed loop.

ANSWER: PAGE 100

▼ DNA

This string of DNA consisting of 16 separate but joined parts has been split and the 16 parts can be rearranged into two identical closed rings.

Can you copy, cut out, and recreate the two identical clones?

ANSWER: PAGE 100

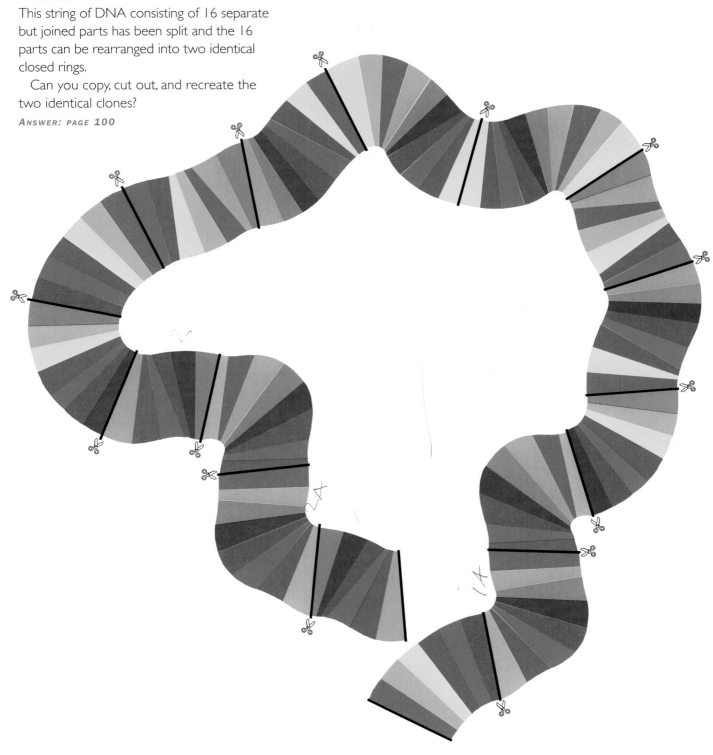

Concentric circles have the same center point. They might not seem very esthetically pleasing at first, but the renowned Russian artist Wassily Kandinsky (1866–1944) created modern artworks such as "Squares with Concentric Circles" to great acclaim.

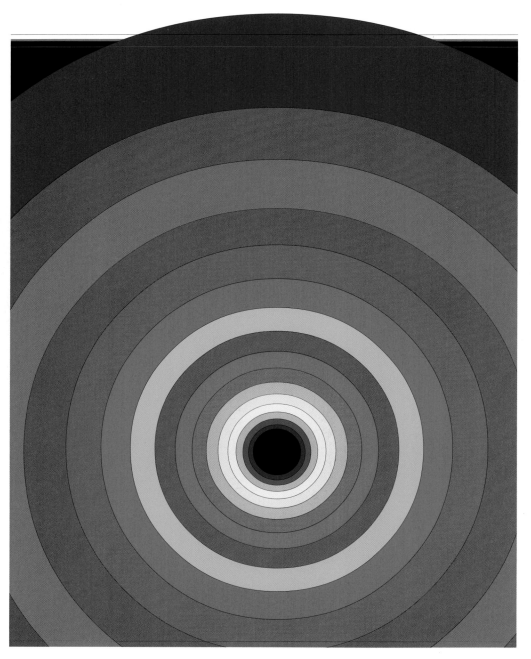

◄ TO INFINITY... THEN STOP

The curvature of a circle's circumference decreases as the size of the circle increases.

What is an infinite circle?

ANSWER: PAGE 101

▼ HYPNOTIC GAZE

Can you estimate the areas of the ten concentric circles below in terms of the unit area of the central circle?

ANSWER: PAGE 101

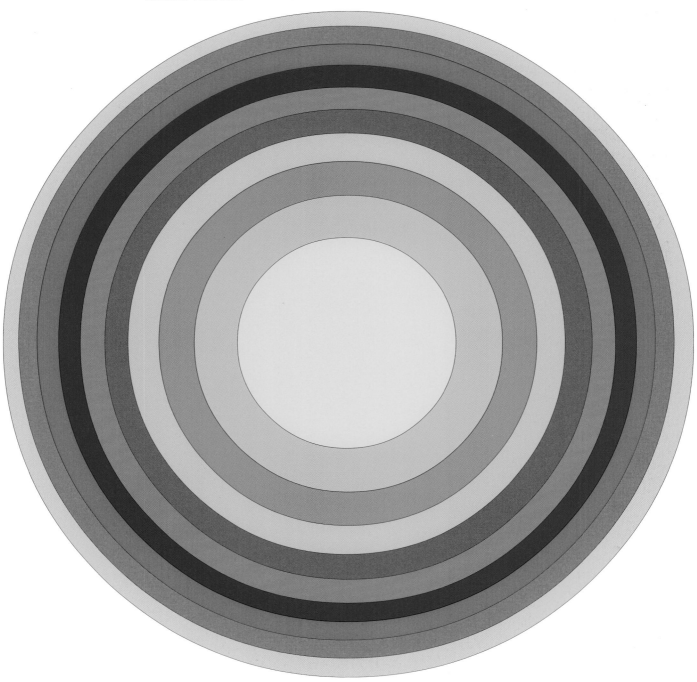

If you used a pair of scissors to enable you to play the puzzles on the previous pages, did you realize you were using a linkage? Maybe not, but read on to discover how linkages can be useful physical models for mathematicians.

✳ Lines and linkages—how to create a straight line

The dimensions of space begin with a point. "A point," explained Euclid, "is position without size." A mathematical abstraction, whose dimensions are zero. The tiniest imaginable portion of space.

The first dimension begins with a line. To build a line, observe the path of a moving point. A single number suffices to locate any point within the line—its distance from some other, arbitrarily chosen, point. A space whose points can be specified by one number is one-dimensional.

A line is the idealization of a rigid rod. Problems about linked rods are problems in the geometry of lines. Mechanics is a form of geometry. A moving line traces out a plane: a space of two dimensions, requiring two numbers to specify a single point—how far north, how far east. There is something fascinating about the motion of linkages. You can easily construct simple linkages yourself from cardboard strips joined by fasteners or eyelets.

A linkage in the plane is a system of rods, or lines, connected to each other by movable joints, or fixed to the plane by pivots about which they can turn freely. The pantograph, an ingenious toy that allows drawings to be traced and reproduced at a different scale (see page 83), is an example of just such a linkage.

Given a number of rigid rods, can a linkage be found that will produce, by the motion of one of its points, a straight line? Try pivoting a single rod at one end: How does the free end move? In a circle. Circular motion is easy and natural for linkages. The trick is to construct straight-line motion in the absence of a fixed straight line.

This is not just a theoretical problem in geometry. The natural motion produced by a steam engine is rotary. While it can be converted to straight-line motion by a piston, pistons require bearings, and bearings are subject to wear. A linkage would provide a more satisfactory solution. The first practical solution, devised by James Watt (1736–1819), was only approximate. The true curve of motion of Watt's linkage was an elongated figure 8, a segment of which was close enough to a straight line for Watt's purposes. From strips of card with holes punched in them and joined by eyelets, you can easily create Watt's linkage, and many others.

The first mechanical device to produce exact straight-line motion is something called Peaucellier's linkage (see page 83), invented in 1864. It is based on a general geometrical principle called inversion. Six links, four of which are of equal length, form an "inverter": If a particular point in the linkage follows a curve, then another point follows the inverse curve. The inverse curve to a straight-line is a circle. A final, seventh, link constrains one of the points to a circle; the other is forced to follow its inverse, the straight line. By a single general insight, the impossible is thus converted to the familiar.

A moving plane generates a space of three dimensions, a solid, requiring three numbers to specify any point. By allowing linkage to move out of the plane and into space, a new variety of forms can be created.

SCHEMATIC REPRESENTATION OF WATT'S LINKAGE When the two rotating rods (blue and green) are of equal lengths, the midpoint of the red linkage (the black point) traces a figure eight pattern over the full journey of the mechanism. The middle part, shown, is a good approximation of a straight line.

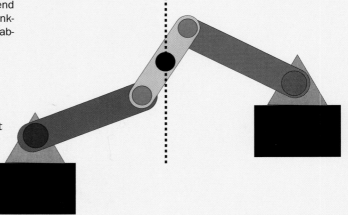

▶ CIRCULAR CALCULATION

Early computers could not draw circles at all, only lines and polygons. (Even today, they are not naturally able to handle curves easily.) Can you describe a way in which an early computer capable of drawing only lines could represent a passable attempt at a circle? One method is hinted by the diagram on the right.

ANSWER: PAGE 101

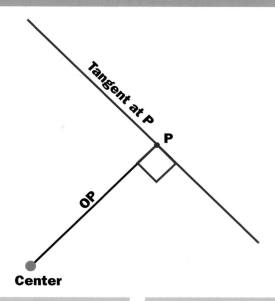

Tangent at P

P

OP

Center

✳ Pythagorean theorem

In this dynamic demonstration model, the Pythagorean theorem is demonstrated by the passage of colored liquid from compartment to compartment (taken from a science museum exhibit).

Mathematicians and scientists use "models" (see box) to give shape to their thoughts and also to help them prove, and sometimes disprove, their theories. These physical manifestations of concepts and theorems are often beautiful objects in themselves and suggest the richness and variety of mathematical expression. They show the relationships between lines, curves, and surfaces, displaying the inherent beauty of pure mathematics. Such analogies are quite different from the real thing, so what can their value be?

There are good answers to this question. Such models can help us understand the basic principles involved, to think about problems creatively, and suggest ideas for deeper understanding and further investigation.

So, if a problem suggests visualization, create a drawing or a sketch of it, or a three-dimensional model or analogy.

✳ Models

Models and analogies are very important in the study of geometry, as well as in science in general, and their value and limitations should be carefully considered in every specific case. Sometimes they are crucial in getting a basic understanding of the problem in question.

One category of models are scale models, which are merely "reduced imitations of the real thing"—as one woman referred to her "model husband!"

Other models are analogies, like using small beads to help you picture the behavior of gas molecules, or flowing liquid (see left) to demonstrate the validity of the Pythagorean theorem for a specific triangle.

Why does it never seem possible to drive from A to B in an automobile? After working out the calculations in these puzzles, you'll better appreciate that road networks can't offer everyone the most direct route to work...

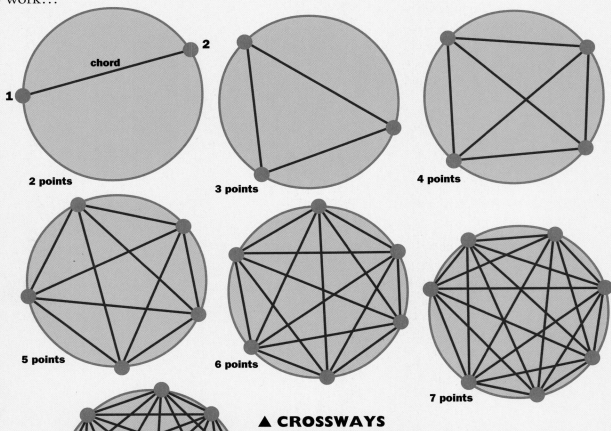

2 points

3 points

4 points

5 points

6 points

7 points

8 points

▲ CROSSWAYS

A line joining two points on the circumference of a circle is called a chord.

How many chords are there in circles with "n" points distributed on their circumferences?

Circles from 2 to 8 points are demonstrated here.

Can you work out the general formula and fill in the table to show the number of chords for "n" from 3 to 20 points?

ANSWER: PAGE 102

$[n \times (n-1)]/2$

Number of points	1	2	3	4	5	6	7	8	9	10	11	12	13	14	15	16	17	18	19	20
Number of chords	0	1	3	6	10	15														

▶ MYSTIC ROSE—19 POINTS

Nineteen points are equally spaced around a circle. Each point is joined to every other point by a straight line.

Can this pattern be drawn in a continuous manner without lifting pencil from paper or retracing any lines?

ANSWER: PAGE 102

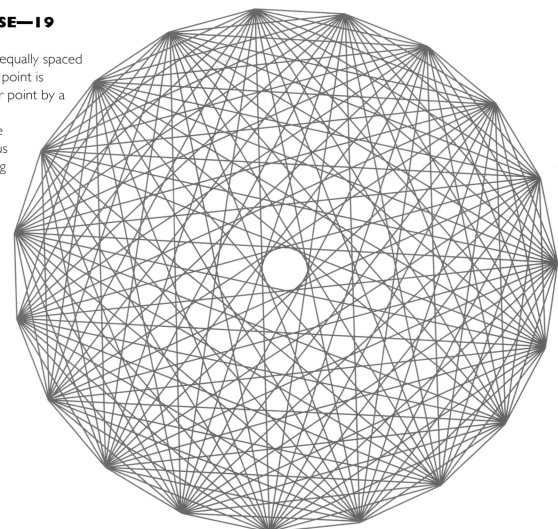

✳ Mystic roses

To create a mystic rose, a set of points is evenly spaced along the circumference of a circle; each point is then connected to every other point by a straight line. It follows that a small number of points leads to the creation of a relatively simple rose. As the number of points increases, however, so too does the complexity of the rose.

In 1809 the French mathematician Louis Poinsot posed the question of the minimum num-ber of continuous lines that would be needed to draw mystic roses of various sizes. (A continuous line is defined as one that is drawn without lifting pen from paper or retracing any of the lines.)

A three-point mystic rose can be drawn with one continuous line, but it is impossible to draw a four-point rose in a continuous line. Two con-tinuous lines would be needed.

Eulerian paths and circuits are concerned with find-ing paths that cover every edge of a graph. Hamiltonian paths and circuits, on the other hand, deal with the problem of visiting all of the vertices of a graph, regardless of whether every line has been traveled across on the way.

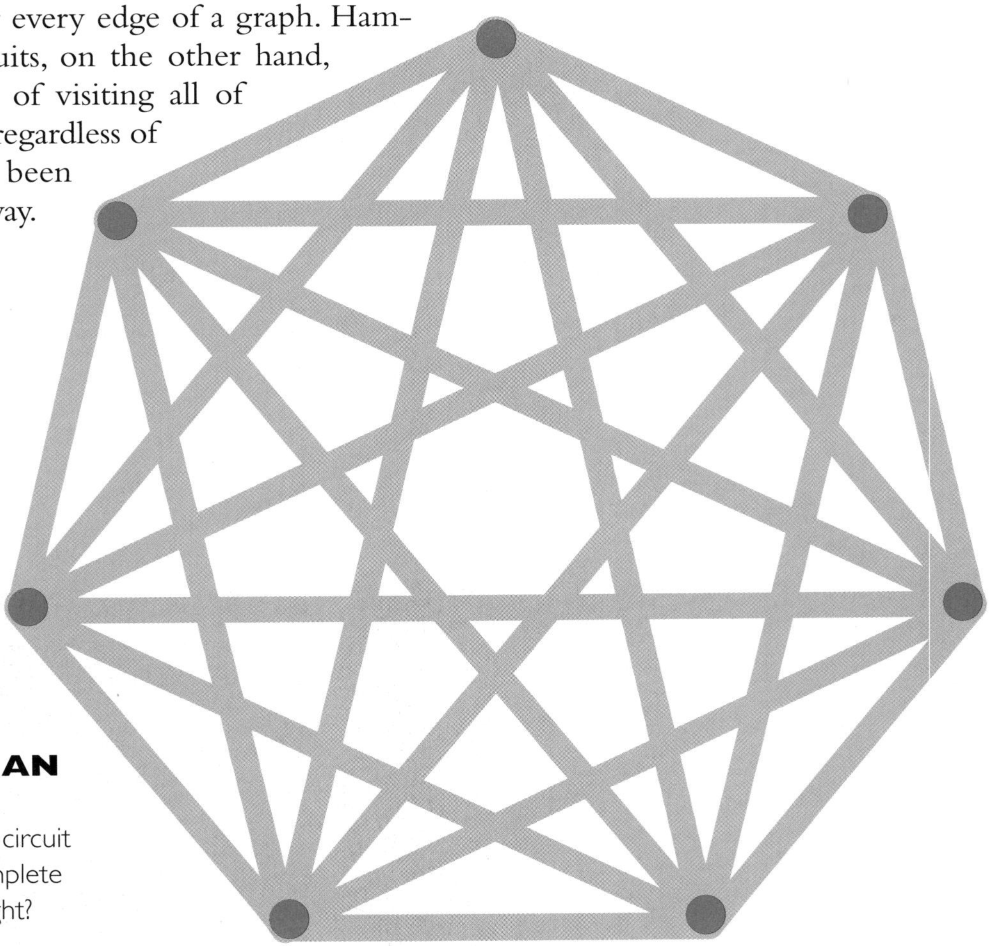

▶THE HAMILTONIAN WAY

Can you find a Hamiltonian circuit (see box, below) in the complete graph on seven points at right?

ANSWER: PAGE 102

❋ Hamiltonian paths and circuits

The Irish mathematician Sir William Rowan Hamilton was especially interested in problems which involved finding a circuit that goes through every vertex exactly once and returns to the starting vertex. He was the first to study these types of problems, which is why they are today called Hamiltonian circuits. Such paths that don't return to the starting vertex after visiting every vertex are called Hamiltonian paths.

Unlike Eulerian paths and circuits, there is no quick method for determining whether a graph has a Hamiltonian path or circuit in general. There are only rules about particular graphs.

For example, if a graph has at least three vertices, and each vertex is connected to more that half of the other vertices in the graph, then it is always the case that the graph must have a Hamiltonian circuit.

▶ BACK TO SQUARE ONE

Can you find a Hamiltonian circuit in this given network, right, starting at the green circle, then moving on to the middle red circle and back? If not, what is the minimal number of edges you need to retrace to achieve the objective?

ANSWER: PAGE 103

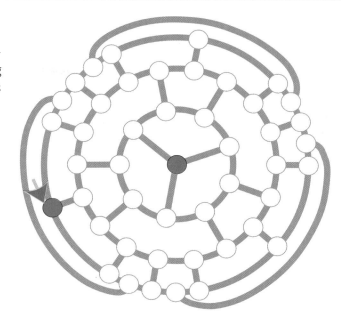

▶THE ICOSIAN GAME

So far we have been traversing graphs on the plane. A more difficult group of problems involves finding routes on three-dimensional objects. One of the first, and still a classic, was invented by Hamilton in 1859, involving a dodecahedron.

He asked whether a route along the edges exists that would come back to the starting point, after visiting all 20 vertices (a Hamiltonian circuit, in other words) and without retracing an edge.

To make it easier to solve such 3-D problems, Hamilton used a two-dimensional diagram of the dodecahedron (called a Schlegel diagram), right, which is topologically equivalent to the three-dimensional solid (bottom right).

Hamilton devised a branch of mathematics to solve similar path-tracing problems on three-dimensional solids, called Icosian calculus.

Bonus puzzle: Can you color the edges of the graph with three different colors, so that following the path traced by any two of the three colors gives a Hamiltonian circuit?

ANSWER: PAGE 103

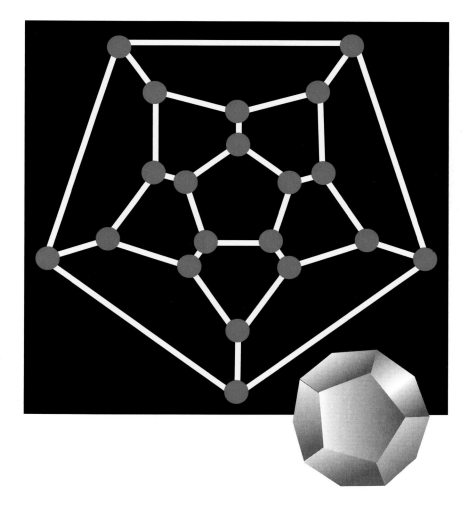

Not to be confused with Hippocrates of Cos, the famous Father of Medicine, the Greek geometer Hippocrates hailed from the Aegean island of Chios. He used a principle of give–and–take and applied it to several forms of curved shapes whose areas could be described in surprisingly simple terms.

▶LUNES OF HIP-POCRATES

Two versions of the Lunes of Hippocrates puzzle are shown here.

Puzzle 1 Can you work out the total of the two areas of the lunes (L and M), in terms of the area of the right-angled triangle A-B-C (A-B is the diameter of the circle)?

Puzzle 2 Hippocrates of Chios discovered this problem while trying to square the circle: A square is inscribed in a circle. Four semicircles are drawn on the four sides of the square, describing four moon-shaped crescents.

Can you determine what the total area of the four crescents will be?

ANSWER: PAGE 104

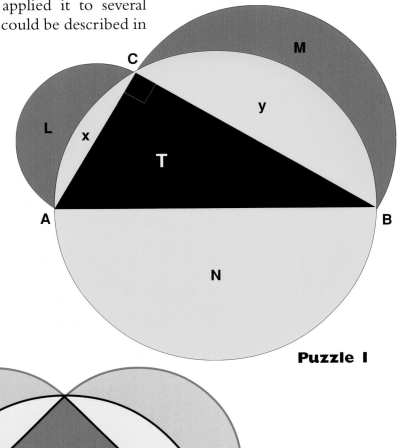

Puzzle 1

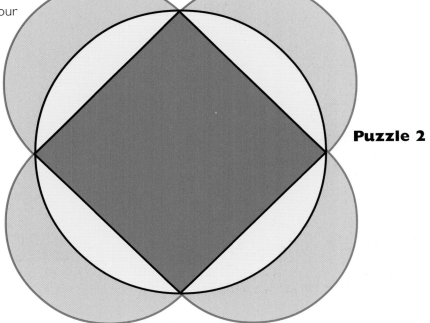

Puzzle 2

▼ THE CIRCLE INSIDE

These four equal squares have sides of 2r length.
Which figure has the largest black region?

ANSWER: PAGE 104

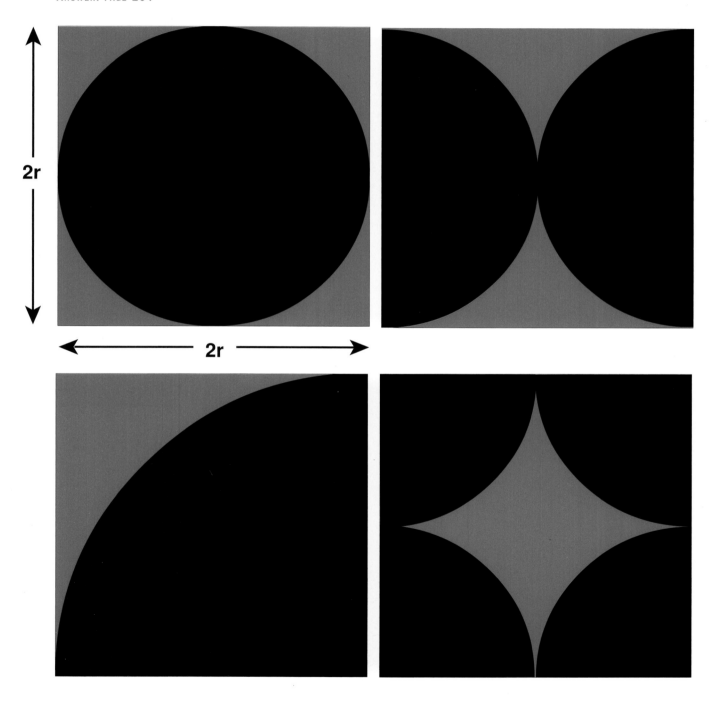

Leonardo was born in the year 1452 in the town of Vinci, in Italy. He was not only a scientist, but an all–round genius, interested in anatomy, art, aviation, architecture, botany, engineering, geology, mechanics, philosophy, and sculpture. He was simply amazing.

"The desire to know is natural to good men.
Leonardo da Vinci

Leonardo's inventions:
1. Scientific studies of flight in birds
2. Parachute
3. Helicopter
4. Experimental device to determine aerodynamic forces
5. Water pumps

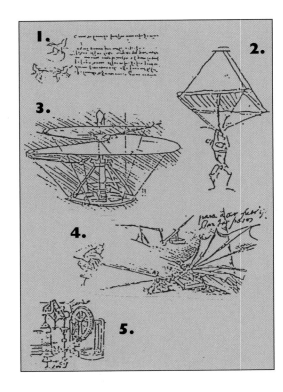

◄ LEONARDO'S DISSECTION

Leonardo da Vinci was fascinated by the problem of dissecting and calculating the areas of curved figures. He gave a few examples of simple dissections involving transformations of curved figures into squares. He drew a more complex curved figure shown above that can be dissected into a rectangle.

Math professor Herbert Wills called the figure a "motif" and found the solution to transform it into a 2-by-1 rectangle. Computer scientist Greg Frederickson in his book **Hinged Dissections** provided a six-piece variation of Wills's dissection.

Can you find a solution?

ANSWER: PAGE 105

❋ Leonardo da Vinci

In all of Leonardo's work he made much practical use of mathematics and geometry. In any serious discussion of those associated with mathematics, science, and art, his name cannot be omitted.

It is said that he could produce a sketch with one hand while making notes with the other. His reputation as an artist dwarfed his fame as a scientist and practical mathematician for years. People praised his *Mona Lisa, The Last Supper, Virgin of the Rocks*, and other paintings.

It was not until centuries after his death, however, that his genius was fully appreciated. Here was a man five centuries ahead of his time, someone thinking about airplanes while other people still thought in terms of sailing ships.

Not only that, he was studying the organs of the human body while others were still persecuting witches and sorcerers. At a time when ignorance and superstition abounded he managed somehow to dissect more than 30 corpses.

Many of Leonardo's ideas never got past the notebook stage. Many others, however, were actually built and used. He designed a self-driven car that ran by springs and traveled at about 15 miles per hour. He designed a set of tracks on which it would run and he put flanges on the wheels so that it could go around curves. And his coin-stamper was the first machine to stamp designs on both sides of a coin at once.

Leonardo devised a self-propelled boat and drew sketches of a double-hulled ship that would float even if one hull were pierced.

He made an apartment ventilator—the world's first air-conditioning apparatus. He invented versions of gears, speed drives, and roller bearings. He built machines for the textile industry, grinding mills, oil presses, a printing press, derricks, cranes, pulleys, and automatic saws.

At age 50 he planned the canalization of the city of Florence, which is still in use today. He is the father of the science of hydraulics, and the founder of hydrostatics.

He worked on the principle of a mechanical clock, which was driven by weights in the same way as a grandfather clock. He developed devices for measuring wind velocity and humidity. His "aerial screw" was a type of helicopter—the power was supplied by springs like those in clocks. He flew model airplanes before Columbus discovered America. There seems to have been no limit to the man's genius.

Leonardo died on May 2, 1519—he was undoubtedly the most creative person our world has known.

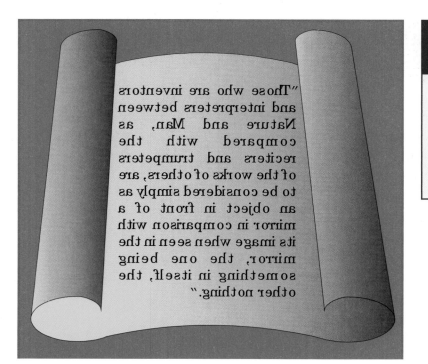

"Those who are inventors and interpreters between Nature and Man, as compared with reciters and trumpeters of the works of others, are to be considered simply as an object in front of a mirror in comparison with its image when seen in the mirror, the one being something in itself, the other nothing."

Leonardo's mirror

Perhaps to disguise some of his work, which was revolutionary, Leonardo made many of his famous notes in mirror writing. That is, he wrote everything backward. You can read his words on the left using a mirror.

(ANSWER: PAGE 105)

In Chinese mythology, the universe was created from a shapeless mass that gradually separated. The lighter elements rose to form Heaven and the force of "yang," while the heavier material clumped together to form Earth, giving rise to the "yin" force. The puzzles on these pages are based on the yin-yang symbol and similar circular constructions.

▼ PRETTY PETALS

This square has sides of 2 units.

Can you find the total area of the four red petals within it?

ANSWER: PAGE 106

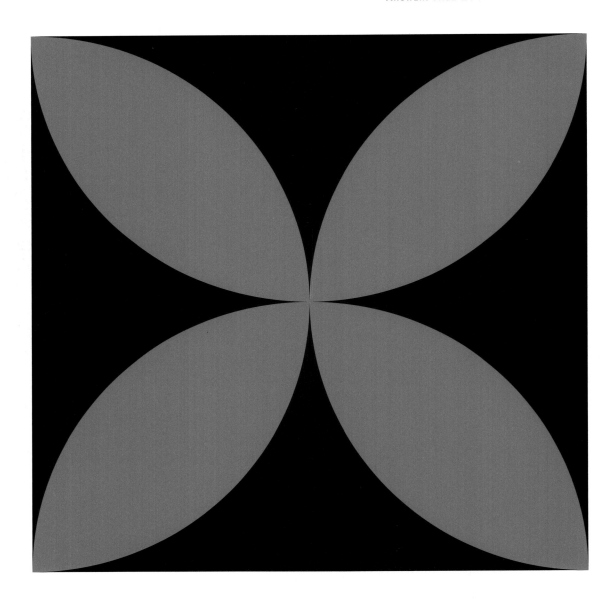

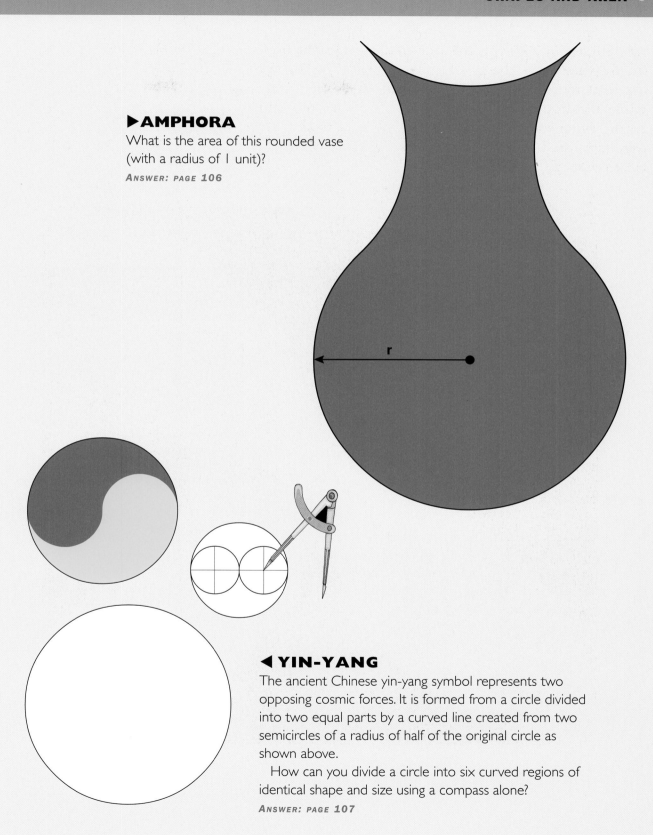

▶AMPHORA

What is the area of this rounded vase
(with a radius of 1 unit)?

ANSWER: PAGE 106

◀ YIN-YANG

The ancient Chinese yin-yang symbol represents two
opposing cosmic forces. It is formed from a circle divided
into two equal parts by a curved line created from two
semicircles of a radius of half of the original circle as
shown above.

   How can you divide a circle into six curved regions of
identical shape and size using a compass alone?

ANSWER: PAGE 107

The early Libyan mathematician Eratosthenes was the first person to produce an accurate measurement of the circumference of the Earth.

489 miles (787 km)

7.2°

▲ MAN OF THE EARTH

Although early Greek geometers made huge theoretical advances, the mathematician Eratosthenes accomplished perhaps the greatest practical achievement. He learned that on a day in midsummer in the town of Syene (near present-day Aswan), the reflection of the noonday sun was visible on the water of a deep well. For that to occur, the sun had to be directly overhead, with its rays pointed directly toward the center of the Earth. On the same day, the noonday sun cast a shadow in Alexandria at an angle of 7.2 degrees, or about $\frac{1}{50}$ of a full circle.

Eratosthenes also knew the north to south distance between Alexandria and Syene, which is about 489 miles (787 km). This information was sufficient for him to calculate the circumference of the Earth with astonishing accuracy.

Can you come to the same result as Eratosthenes and calculate, as he did, the circumference of the Earth?

ANSWER: PAGE 107

ERATOSTHENES (276 B.C.–194 B.C.)

Eratosthenes was born in Cyrene, now in Libya. He studied in Athens and later became the third librarian at Alexandria, in a temple of the Muses. Called the Mouseion, it contained hundreds of thousands of papyrus and vellum scrolls.

Nicknamed Beta by his contemporaries as he fell short of the highest ranks in many categories of his work, he is now recognized as a great scholar in many disciplines. His accomplishments are today considered not only historically important, but also remarkable for providing early examples of modern scientific methods.

One of his most important works, Platonicus, deals with the mathematics underlying Plato's philosophy.

In addition to his surprisingly accurate measurement of the Earth's circumference, he ascertained the distances to the sun and moon, using data during lunar eclipses. He also measured the tilt of the Earth's axis with great accuracy, obtaining the value of 23° 5' 15". Furthermore, he produced a calendar that included leap years.

It has been recorded that he committed suicide on becoming blind in old age.

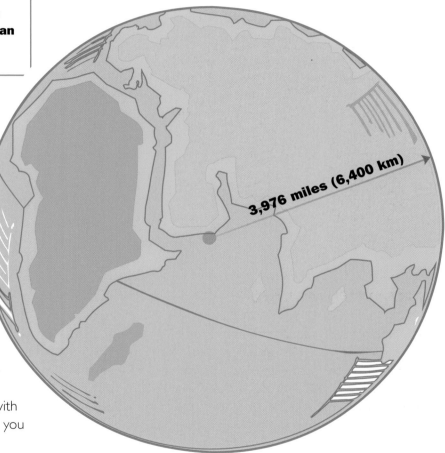

? DID YOU KNOW?

The first attempt to communicate with beings on the moon or on Mars began more than 150 years ago. The German mathematician Carl F. Gauss (1777–1855) suggested that there be erected in Siberia a giant figure of Euclid's demonstration of the Pythagorean theorem.

My conviction is that the Earth is a round body in the corner of the heavens.

Plato's Phaedo, attributed to Socrates, the earliest known written record of the idea that the world is round

▶ **WHAT ON EARTH?**

Archimedes is said to have been so proud of his discovery of the formulas for calculating the surface area and volume of a sphere, from its relationship to the surface area and volume of a cylinder of the same diameter and same height as the diameter of the sphere, that he wanted the discovery to be memorialized on his gravestone (see right).

Idealizing Earth as a perfect sphere with a radius of 3,976 miles (6,400 km), can you find the surface area of the Earth?

ANSWER: PAGE 108

3,976 miles (6,400 km)

Every cylinder whose base is the greatest circle in a sphere and whose height is equal to the diameter of the sphere has a surface area, including the bases, ³⁄₂ the surface of the sphere, and its volmume is ³⁄₂ of the volume of the sphere.

Archimedes

The area of math called calculus is exceptionally useful when considering 3-D shapes. If you know the volume of a simple object, such as a sphere, you can work out the surface area (and vice versa). But luckily, just some high school math is all that is needed here.

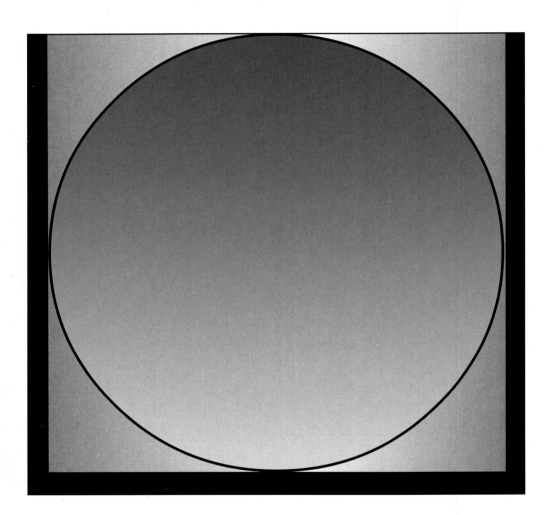

▲ HYDROSPHERE

You have before you a thin-walled glass sphere filled with water fitting exactly into a cylindrical box, the width and height of which match the diameter of the sphere

If you break the glass sphere and pour the water into the cylinder, how much of the cylinder's volume will be filled with water?

ANSWER: PAGE 108

DaVinci
THE EXHIBITION

IMAGINE
EXHIBITIONS GALLERY

Thu, May 22, 2014
10am - 9pm

*A hands-on exploration of
the life and work of the
original Renaissance man,
Leonardo da Vinci!*

<u>NO RE-ENTRY</u>

SERVICE FEE INCLUDED
NO REFUNDS OR EXCHANGES

GRANDE
EXHIBITIONS   **imagine**

PRINTED   ACCT#
                           11548703

ADMIT ONE
Ticket stub will be
detached upon entry.

GR-ENT-0600EA-05-BB

The person using the license rights granted by this ticket ("holder") agrees to the following by using the ticket: Holder voluntarily (1) assumes all risks of injury or property damage arising from or related to, or occurring at the site of, the event the subject of this ticket whether occurring prior to, during, or subsequent to the actual event and (2) releases everyone connected in any way to the event including The Venetian and The Palazzo, and all of their parents, subsidiaries, affiliates, and any of their agents, officers, directors, owners, shareholders, and employees (collectively, the "released parties") from any claims arising from or related to the event even if any of the released parties are negligent.

TICKET MAY BE REVOKED OR VOID: This ticket is a revocable license and admission may be refused for any reason or no reason upon refunding of ticket face amount. This license is governed by Nevada law and any action brought regarding the use of or in any way related to this ticket, shall be brought in Clark County, Nevada. Tickets obtained from unauthorized sources may be lost, stolen or counterfeit, and, if so, are void. It is unlawful to reproduce this ticket in any form. Damaged, torn, disfigured and/or detached tickets are void. Management reserves the right, without the refund of any portion of the ticket purchase price, to refuse admission to or eject any person who is deemed by management to be (1) acting disorderly, (2) using vulgar or abusive language, or (3) not complying with any house rule, including the rules on this ticket. Any activity deemed likely to interfere with the performance or endanger persons or the facilities will result in immediate eviction from the venue without refund of ticket price.

NO REFUNDS, CANCELLATIONS, AND DATE CHANGES:
There shall be no refunds or exchanges, except as described on this ticket or required by law. Performance dates and times subject to change without notice. Any claim for a cancelled performance for which there is no rescheduled date shall be limited to refund of ticket face amount. Such claim must be filed with seller within 30 days after performance was to have occurred. This ticket cannot be replaced if lost, stolen or destroyed, and is valid only for the performance and seat for which it was issued. Unless indicated otherwise, price includes all applicable taxes.

RESALE AND PROMOTION RESTRICTIONS; BOX OFFICE PURCHASES:
Resale of this ticket may be prohibited under certain laws and circumstances. These restrictions include, but are not limited to, the following: under Nevada law, NRS § 597.830, it shall be a misdemeanor to add to the price of the admission or ticket more than the actual amount of any federal or state tax thereupon imposed. This ticket shall not be resold on the premises except through an authorized box office agent. Ticket may not be used for promotion or other trade purposes, in advertising, sweepstakes, or other programs or materials without the prior written consent of promoter and venue. Unlawful resale or attempted resale is grounds for seizure and cancellation without compensation.

VENUE RULES; RECORDING AND EXHIBITION RIGHTS:
Certain items may not be brought into the premises, including without limitation, alcoholic beverages, bottles, cans, containers, illegal drugs, controlled substances, food, signs, fire arms, weapons, other items deemed likely to endanger persons, cameras, recording devices, laser devices, bundles, and/or containers of any kind. Holder hereby consents to the reasonable inspection of his/her person or possessions for any such item and to the confiscation thereof without compensation. Holder agrees not to transmit or aid in transmitting any description, account, picture, or reproduction of the production, performance, exhibition or event for which this ticket is issued. Holder shall not be permitted to distribute literature or other materials or to sell or distribute merchandise in the theatre or on the premises. Holder acknowledges that the performance may be recorded, broadcast, or otherwise publicized or exhibited, and hereby grants permission in perpetuity to the use of Holder's image or likeness, in whole or in part, in connection with any video display, filming, transmission, recording of the performance or any subsequent publication, display or performance of any of the foregoing.

Late seating subject to management discretion. Holder is free to exit theperformance at anytime, however, re-entry will be controlled and only allowed at certain intervals. Ticket stub necessary for re-admission to performance. No smoking in theatre. Special fog and strobe effects used during the performance.

All persons, regardless of age, must have a ticket for admission.
Management reserves all rights.

THE VENETIAN® | THE PALAZZO®

THE VENETIAN® | THE PALAZZO®

SALE

Guest: GAY VEIT

Method: Cash

CC #:

Auth #:

Guest: GAY VEIT

Amount: 45.00

Fees: 0

Total: 45.00

Signature: _____

Account #: 11528703

Event Code: DVE40522

Seats:

Print User: CATOB

GR-ENT-00066-08-08

▶ ▼ SPHERE OF INFLUENCE

This time you have a thin-walled glass sphere filled with water fitting exactly into a cubical box, the width and height of which exactly match the diameter of the sphere.

If you break the glass sphere and pour the water into the cube, how much of the cube's volume will be filled with water?

ANSWER: PAGE 108

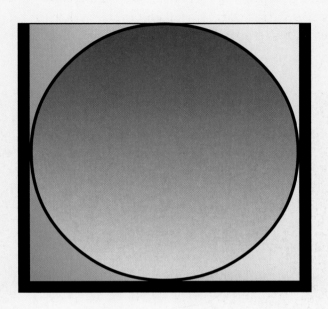

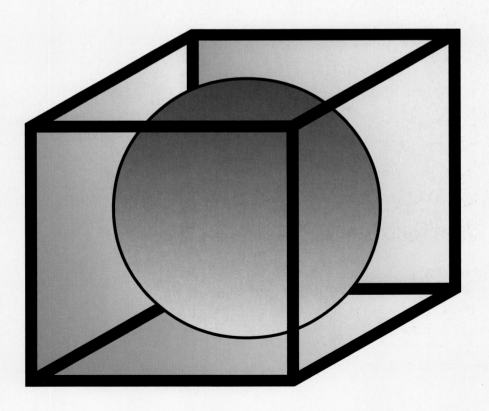

The Italian mathematician Leonardo Fibonacci invented a mathematical sequence (now called the Fibonacci sequence) that was later discovered to present itself in many areas of science, math, and nature. The *Fibonacci Quarterly* is a modern journal devoted entirely to mathematics related to the Fibonacci sequence.

LEONARDO PISANO, AKA FIBONACCI (1170–1250)

Leonardo Pisano, known by his nickname Fibonacci, was probably born in Pisa, Italy, and also died there.

Educated in North Africa, he traveled much with his father, who held a diplomatic post. Returning to Pisa around 1200, he wrote a number of important handwritten mathematical texts, the most important of which was Liber Abaci *(1202). Based on mathematical knowledge that Fibonacci had accumulated during his travels, its most important contribution was introducing the Hindu-Arabic place-value decimal system and the use of Arabic numerals into Europe. One section contains a large collection of problems aimed at merchants. Another section is devoted to the introduction of the Fibonacci numbers and the Fibonacci number sequence, a contribution for which he will always be remembered.*

Fibonacci rabbits

The most famous recreational mathematics problem concerning number sequences is this classic from 1202, seen on page 33.

This rabbit-breeding puzzle was found in Fibonacci's book Liber Abaci.

Fibonacci hypothetically assumed that every pair referred to is composed of a male and a female rabbit and that they bear young two months after their birth, when in reality they reach maturity only after a period of four months.

▼ RAMPANT RABBITS

Each rabbit represents a pair (a male and a female) and each pair bears
a pair of young two months after their birth, and bears another pair
of young every month after that. How many are produced in a single
year? Note: The original pair were born in the previous December.

ANSWER: PAGE 109

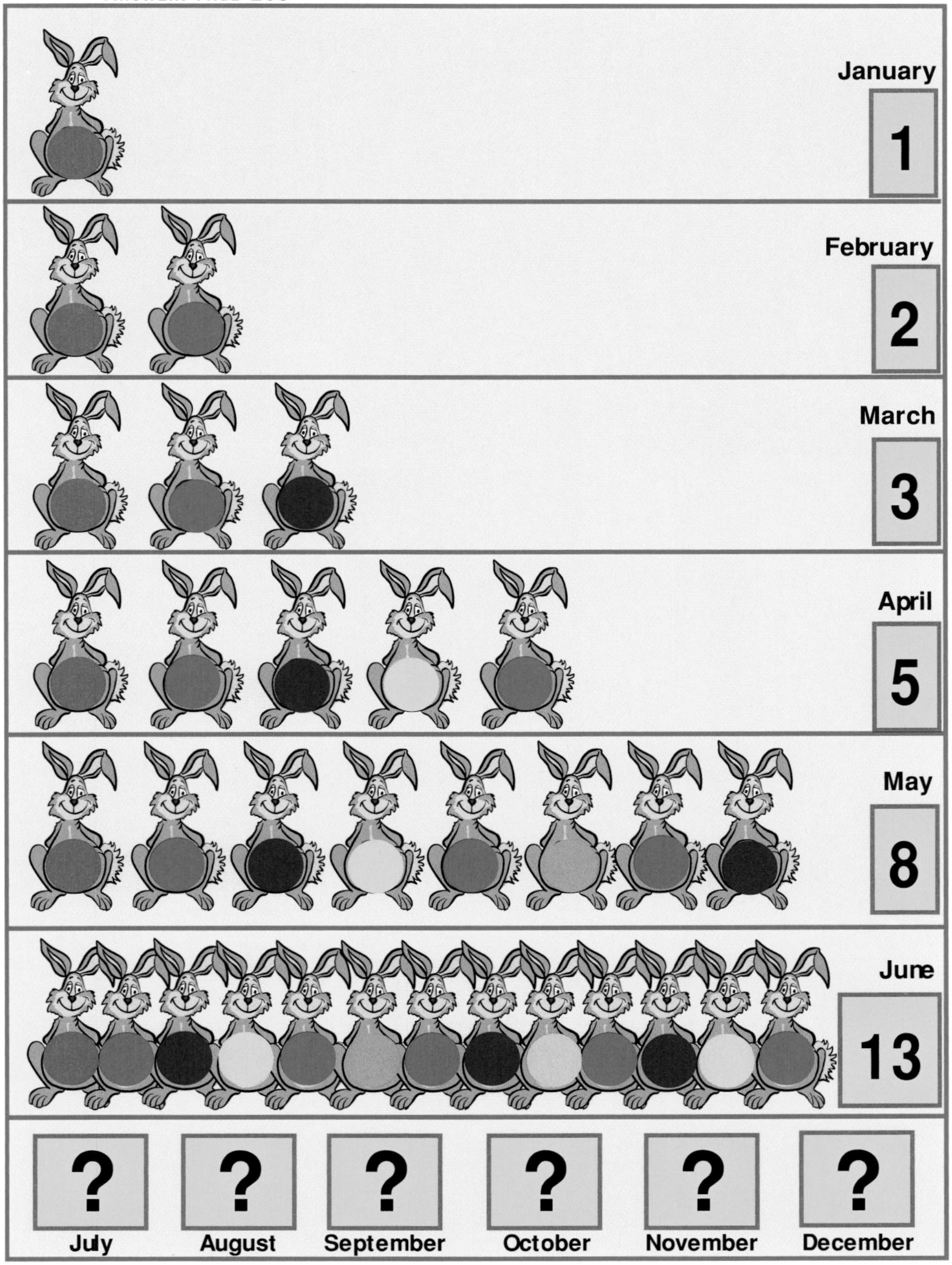

January	1
February	2
March	3
April	5
May	8
June	13
July	?
August	?
September	?
October	?
November	?
December	?

▶ SHOWY SUNFLOWER

Leonardo Fibonacci discovered the number sequence on the left as a recreational mathematics exercise.

Can you find which of the two numbers are represented in the sunflower? Can you also discover the secret of the sequence and continue its terms as far as you like?

ANSWER: PAGE 109

❓ DID YOU KNOW?

The study of leaf positioning around a stalk has a special name. It is called phyllotaxis.

Leaves

Mathematical forms are present in plant life. Even millions of years ago, long before Fibonacci was born (see page 32), certain leaves were arranging themselves around stalks in obedience to his series.

Many plants take ⁹⁄ᵇ of a turn of a circle between leaves, where "a" and "b" are Fibonacci numbers, for example, ²⁄₅ for an apple tree. This method "spreads" the leaves well and prevents them from competing for sunlight.

We know today the economic law involved, but how did the plants come to know it? Was it the result of countless experiments in search of maximum efficiency, culminating in a plant habit, or was some other law involved that only the future will reveal?

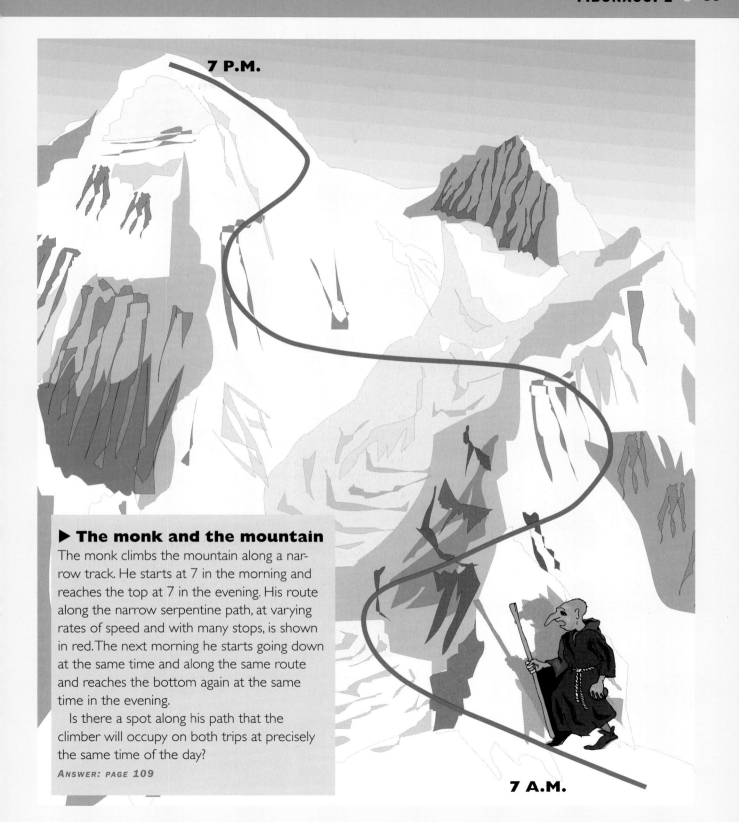

7 P.M.

▶ The monk and the mountain

The monk climbs the mountain along a narrow track. He starts at 7 in the morning and reaches the top at 7 in the evening. His route along the narrow serpentine path, at varying rates of speed and with many stops, is shown in red. The next morning he starts going down at the same time and along the same route and reaches the bottom again at the same time in the evening.

Is there a spot along his path that the climber will occupy on both trips at precisely the same time of the day?

ANSWER: PAGE 109

7 A.M.

A sphere is perhaps the simplest solid shape that one can imagine. It has no corners or edges. Every spot on the outside of a ball is exactly the same distance from the center as every other spot. In the same way, the Earth, moon, planets, and the sun are round. Of course, the matter is not quite so simple. Other forces distort planets and alter their spherical shapes.

✳ Packing circles and spheres

The German astronomer Johannes Kepler found that there are two ways to arrange spheres regularly in a plane: in a square lattice or in a hexagonal lattice (like in a honeycomb). These two arrangements can be stacked in space in several ways. Square layers can be stacked so that the spheres are vertically above each other, or the spheres in each layer nestle into the gaps between four spheres in the layer below.

With hexagonal layers there are also two possibilities, aligned or staggered, as shown at the bottom of the next page. If the spheres in these arrangements are allowed to expand, they form different three-dimensional shapes. You can see this easily when trying to arrange oranges into a pyramid. Depending on whether you use a square or triangular base, the packing will be different.

The efficiency of a packing system is measured by a number, its density (that is, the proportion of space that is filled with spheres).

The problem of close sphere-packing is closely related to geometric solids that can be fitted together to fill space completely. Kepler tried to obtain such solids by imagining each packing sphere expanding to fill the intermediate space.

Kepler's insight found a fundamental connection between the formation of a snowflake, the construction of a honeycomb by bees, and the growth of a pomegranate. According to his theory, the regular, symmetric patterns that arise in each case can be described and explained in terms of "space-filling" geometric figures, such as his own discovery, the rhombic dodecahedron, a solid figure with twelve identical rhombic faces.

The rhombic dodecahedron, which can be packed to fill space, is the key to the sphere-packing problem (we'll be meeting one later in the answer section of the book).

▼ PACKING DISKS OR CIRCLES

There are two ways of filling the plane with circles as shown.

Can you find the percentage of the total areas covered by rectangular and hexagonal packings?

ANSWER: PAGE 109

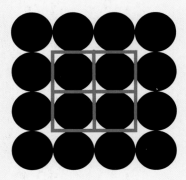

Rectangular (square) packing

Hexagonal (triangular) packing

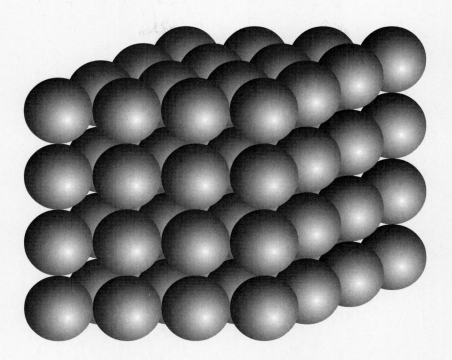

CUBIC SPHERE-PACKING
In square layers, corresponding spheres are stacked vertically above each other.

HEXAGONAL SPHERE-PACKINGS
There are two ways to add a hexagonally packed layer.
Right: Spheres in the third layer are directly above those in the first.
Below: Face-centered cubic packing.

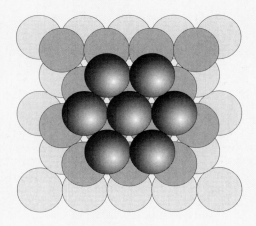

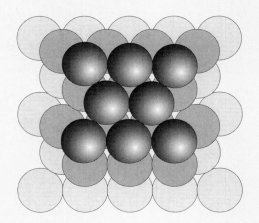

Circle (and sphere) packing, once an obscure mathematical recreation, has become an indispensable tool to modern science, especially in problems where cramming things together into the tightest possible arrangement is essential. The mathematics of close-packed spheres is important, especially in the study of crystals and atomic structure.

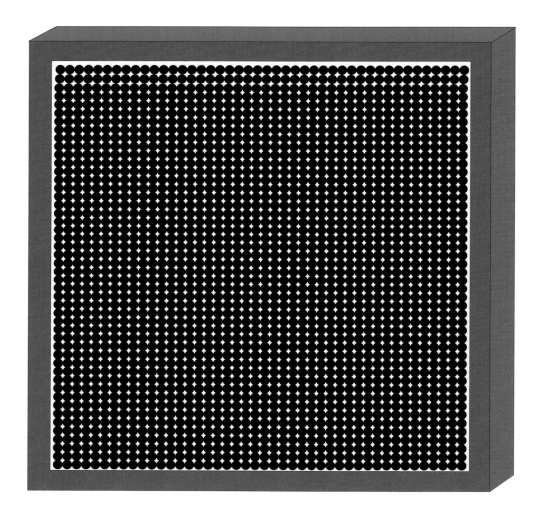

▲ BLACK BALLS

A hermetically sealed box in which are enclosed thousands of small identical steel balls is shown. What will happen each time the box is shaken?

ANSWER: PAGE 110

> **M**en are never more ingenious than when they are inventing games; the mind finds satisfaction in this activity.
>
> *Leibniz (1646–1716)*

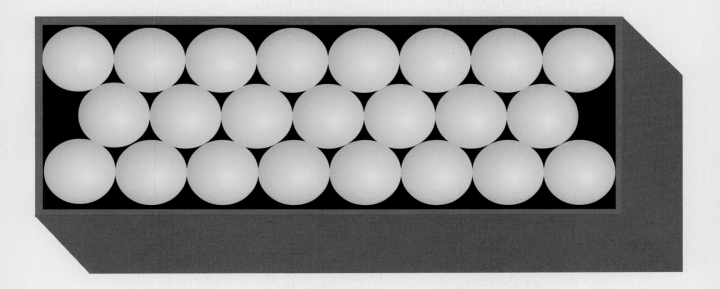

▲ PACK IT IN!

There are 23 golden spheres in this rectangular chest, in a tight-fitting arrangement. How many spheres can be removed from the chest so that the remainder still produce a tight fit? (Tight in this sense means that each sphere is securely held by adjacent touching spheres, so that it can't be moved from its place.)

ANSWER: PAGE 111

▼ BOTTLE IT UP

You have a crate in which 48 identical bottles of wine are tightly packed.

Can you repack them to make space for additional bottles of the same size? How many more bottles can you pack in your crate?

ANSWER: PAGE 111

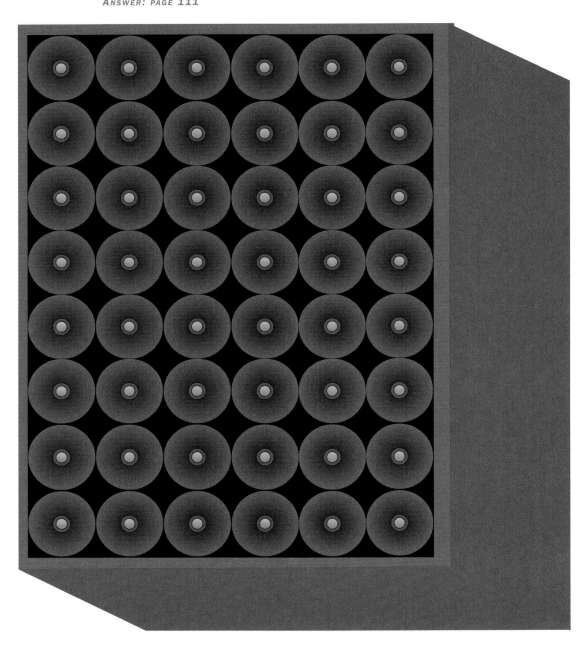

PACKING CIRCLES

Here you can see the densest packings of two to ten circles into a unit square. The numbers in bold are the diameters of the packing circles in terms of the side of the unit square.

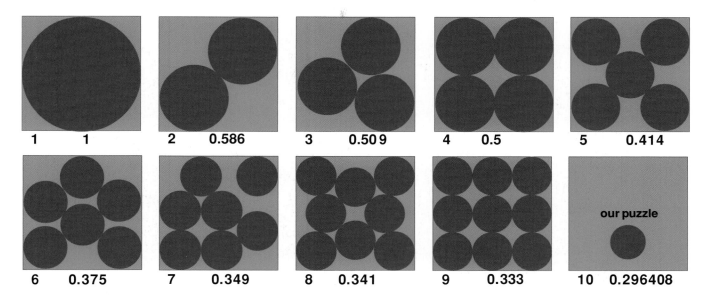

1　**1**	**2**　**0.586**	**3**　**0.509**	**4**　**0.5**	**5**　**0.414**
6　**0.375**	**7**　**0.349**	**8**　**0.341**	**9**　**0.333**	**10**　**0.296408**

our puzzle

► PACKING CIRCLES: TEN IN A SQUARE

Keeping in mind the examples above, can you fit ten identical circles into the red square box, right?

Overlapping of the circles is not allowed. Neither is a circle allowed to leave the red square area.

You can play this puzzle using a dime. If you use any other coin ensure that the square you draw is 3.374 times larger than the coin's diameter.

ANSWER: PAGE 112

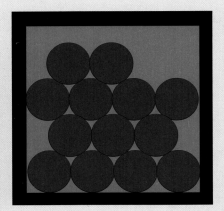

◄ PACKING CIRCLES: 13 IN A SQUARE

We mentioned on page 36 that hexagonal packing is the densest possible packing, covering just over 90% of the area by circles, and proven by Thue's Theorem (named for the Norwegian mathematician Axel Thue).

But this is strictly true only if there is an infinite amount of space.

When space is limited, the situation is completely different as we have seen before.

For example, you can't fit nine circles into a smaller square than a 3-by-3 square, no matter how you try.

On the other hand, 13 circles packed in a hexagonal packing in a 4-by-4 square is not the best packing.

Can you do better and fit 13 circles into the smallest possible square, shown on the left? You can play this puzzle using 13 dimes. If you use another coin make sure that the square you draw is 3.732 times larger than the coin's diameter.

ANSWER: PAGE 112

100 circles

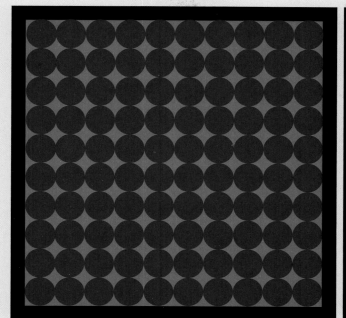

10-unit sided square

105 circles

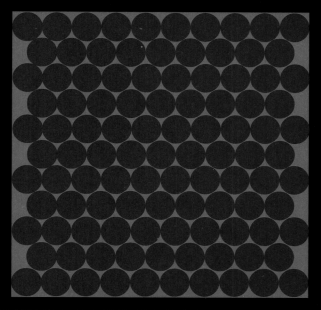

10-unit sided square

10-unit sided square

▲ PACKING CIRCLES: 100 IN A SQUARE

If the diameter of the packing circles is 1 unit, you can easily pack 100 circles in a square with sides of 10 units.

You can do better by packing the circles in a hexagonal array in which you can pack 105 circles as shown.

But can you do even better?

ANSWER: PAGE 113

Suppose you are given a number of identical or non-identical circles to pack inside a larger circle. What is the smallest size that the large circle can be to be able to fit "n" identical or nonidentical circles into it without any overlap?

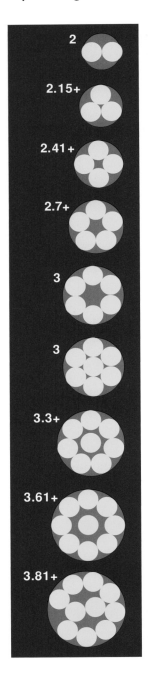

✳ Packing circles in a circle

Newer problems about touching circles involve the packing of "n" identical or nonidentical circles, without overlap, into a specified boundary of the smallest area, circular or other.

Such problems have great practical uses in everyday life. What, you may ask, is the smallest size of cylindrical box into which you can pack milk bottles?

No general solution is known as yet, even when the boundary of the region is as simple as a circle, a square, or an equilateral triangle, and in each case best solutions have been found only for very low values of "n."

When the packing boundary is a circle, proofs are known only for n = 1 through n = 11, first given in 1969. No general solution exists.

In the illustrations on the left, the densest packings of one to ten identical circles onto circles are shown (n = 1 through

n = 10). The numbers beside the unit squares are the minimum diameters of the outside circle with the unit circles inside in each case.

As we've seen on the previous pages, there are many other packing problems, most of which are equally baffling, especially those that allow irregular packing.

It is fairly easy to show that hexagonal lattice packing is the most efficient regular packing of circles. It is enormously harder (though it has been done nonetheless) to show that no irregular packing can be denser.

The analogous problem of spheres packed into space poses more severe problems still. The densest regular packing is known: But whether any irregular packing can do better is still a mystery. At present the answer is suspected to be no, but that remains unproven.

The densest packings of two through ten unit circles into circles. The numbers are the minimum diameters of the outside circle in terms of the inside unit circles.

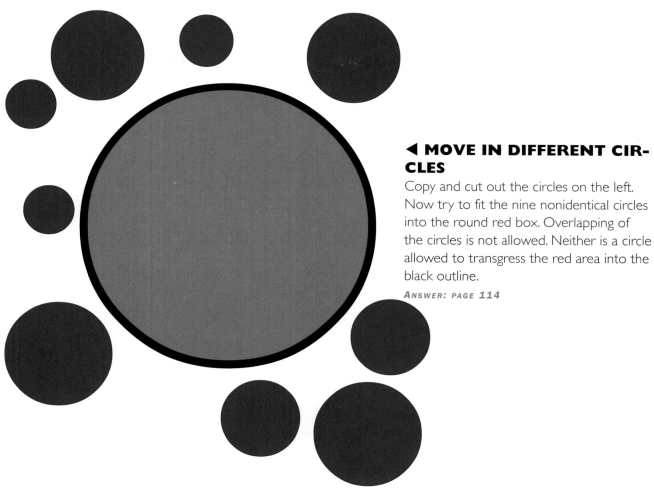

◀ MOVE IN DIFFERENT CIRCLES

Copy and cut out the circles on the left. Now try to fit the nine nonidentical circles into the round red box. Overlapping of the circles is not allowed. Neither is a circle allowed to transgress the red area into the black outline.

ANSWER: PAGE 114

▶ CANNONBALL CLUMP

Viewed from above, the stacked cannon balls balls form hexagonal arrays layered on top of one another.

In 1690, the German astronomer Johannes Kepler considered this the best way of stacking spheres, but only in 1996 was he finally proved to be right. The balls occupy just over 74% of the volume.

Can you guess what would happen to the spheres if you compress the stack together until all the air gaps disappear? In other words, what shape would the spheres form if they were to be pressed together?

ANSWER: PAGE 114

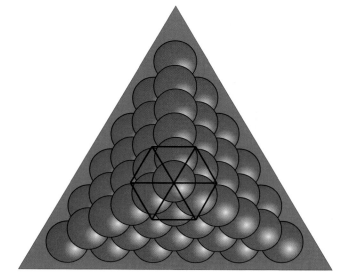

mong modern high-tech toys and games, there is still a prominent place left to the world's oldest word-processing system, which has no memory and no spell-checker. It cannot be upgraded, needs constant maintenance, and is purely analog, but it never crashes. You can even chew it!

✳ Paper-and-pencil games

In this age of the computer and high-tech electronic games, there is an unobtrusive but steady rise in the popularity of quite simple but well-designed basic games that are fun to play.

Paper-and-pencil games and many board games belong to this category. In spite of their seeming simplicity, they are often based on sound mathematical principles, with clearly defined educational objectives. They offer opportunities for fun and enjoyment, combined with a need for logical thinking, strategy, and problem-solving.

▼ THE GAMEBOARD

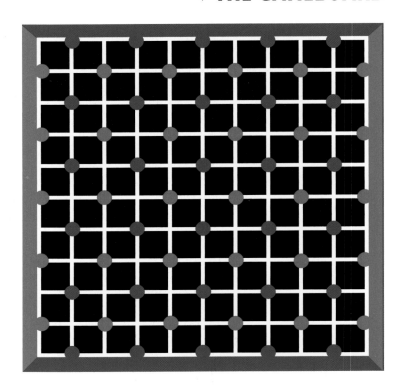

▶ SIDE-TO-SIDE

All you need for this game is graph paper and a pencil. Two players choose colors and alternate moves by connecting two adjacent color points of their color. The first player to create an unbroken line connecting two opposite sides of the gameboard of his or her color is the winner.

In the sample game, blue (the second player) wins the game.

◀ SAMPLE GAME

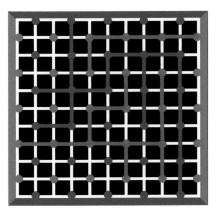

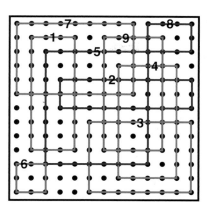

◀ SAMPLE GAME

▼TERRITORY WAR

Each player must complete a rectangle (or square). Lines may cross or meet at a corner, but must not trace over an existing line. The rectangle drawn must enclose at least one "free dot" (a dot that has no line passing through it, or originating at it).

    The player who completes the last possible rectangle is the winner, as shown in the 9th move of the sample game left. Four gameboards are provided below.

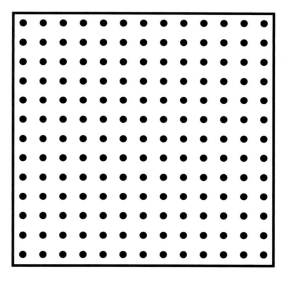

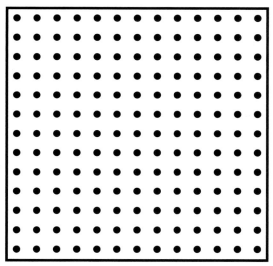

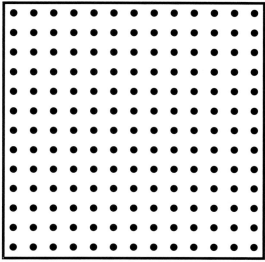

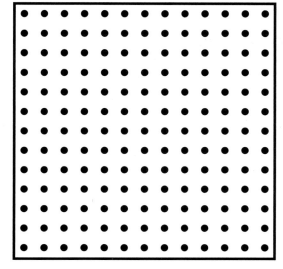

ating back 14 centuries, chess is one of the world's oldest games. But many new games were invented in the 20th century, too, such as the one below by Sid Sackson (1920–2002).

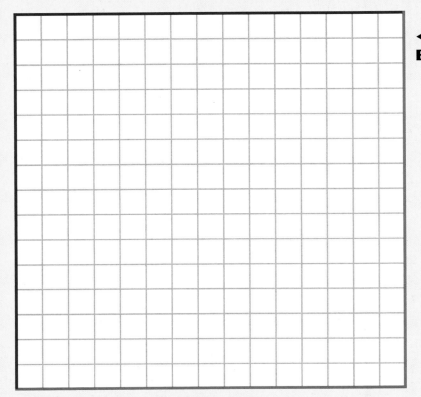

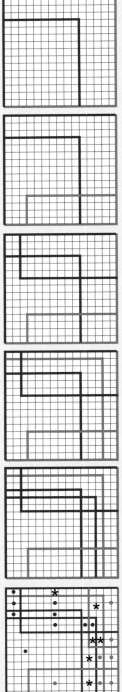

◄ THE GAME-BOARD

▲ CUTTING CORNERS

Two players each play one of the colors (red or blue). They alternate moves by drawing a corner along the square grid (a grid is not actually necessary for playing the game, but we've used ome here to make the games easier to follow). On the first move of the game, at least one of the edges connected must be of the opponent's color. On subsequent moves, each line must either cross a line of the opponent's color or touch an edge of the opponent's color. Each line drawn must cross one more line than the previous one (so the second line in the game crosses one line, the third line crosses two, and so on).

The game ends after each player has drawn three lines. At the end, a section is won by a player if there are more sides of his or her color around it than there are sides of the opponent's color. If the sides are equally divided, it belongs to neither player (*). The sample game won by the blue player on the right will clarify the above rules.

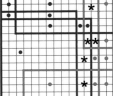

● **9 sections**

● **7 sections**

▲ SAMPLE GAME

▼ ZIGZAG GAME

The object of the zigzag game is to capture the enemy's pieces.

Players choose their colors and place their five playing pieces on the starting lines of the board as shown by the white circles. Players then alternate moves, which can be in any direction—forward, back, left, or right.

On your turn, you can move one piece two spaces if you follow your own color, or one space if you move along a line of your opponent's color.

An opponent's piece is captured by simply moving onto its space, then removing it from the gameboard. One player invariably comes out on top.

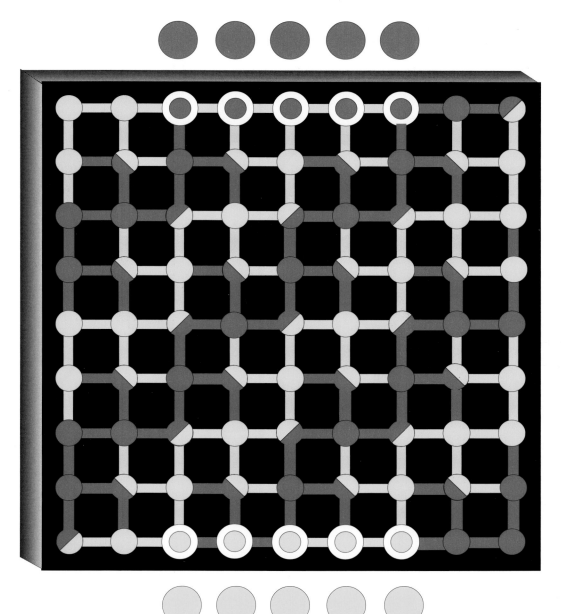

You may be familiar with the infuriating "15" puzzle, where 15 numbered squares are moved around by sliding them one at a time into an empty space so that the numbers are in sequence. In this colorful new twist, we use a hexagonal arrangement.

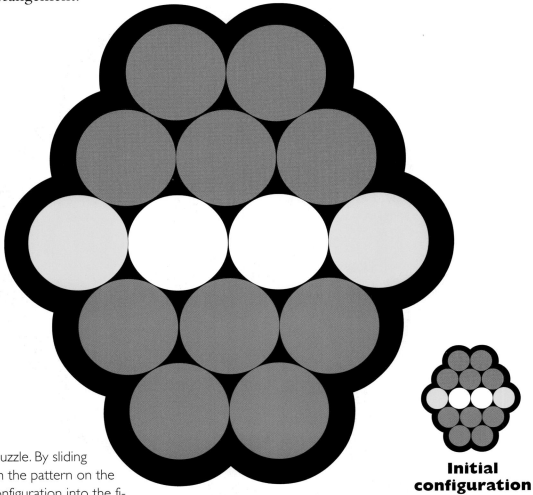

▶ HEXSTEP I

This is a sliding disk puzzle. By sliding moves only, transform the pattern on the left from the initial configuration into the final configuration, as shown. A move consists of sliding a circle into a blank white space. For example, a first possible move could be either moving a middle upper green disk or a middle bottom orange disk into one of the two possible empty spaces, etc.

In how many moves can you achieve the transfer?

ANSWER: PAGE 114

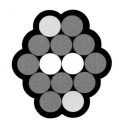

Initial configuration

Final configuration

▼ HEXSTEP 2

A move is defined as sliding a circular disc into one of the white spaces.

Puzzle 1: How many moves will it take you simply to swap the positions of the red and blue discs?

Puzzle 2: How many moves will it take you to get from the initial configuration to the configuration shown bottom right?

ANSWER: PAGE 115

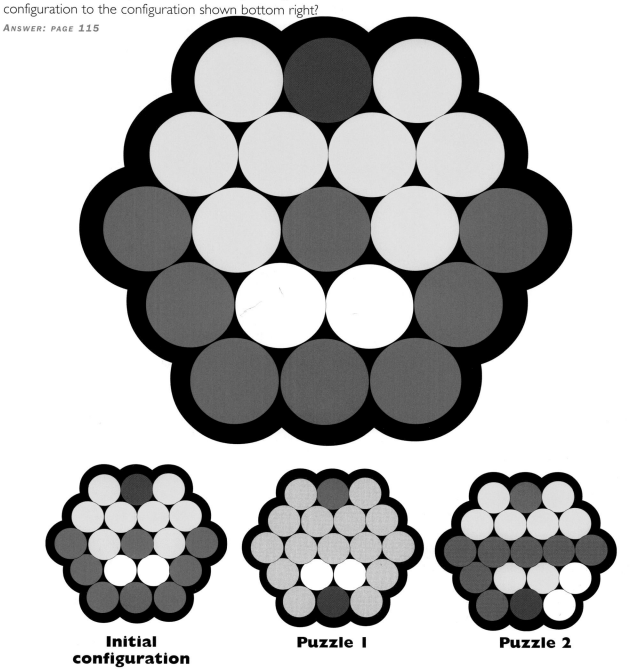

Initial configuration   **Puzzle 1**   **Puzzle 2**

Did you know that dolphins sleep with one eye open, so that they can swim in circles to look out for predators? If the thought of that hasn't made you dizzy, try these circular conundrums.

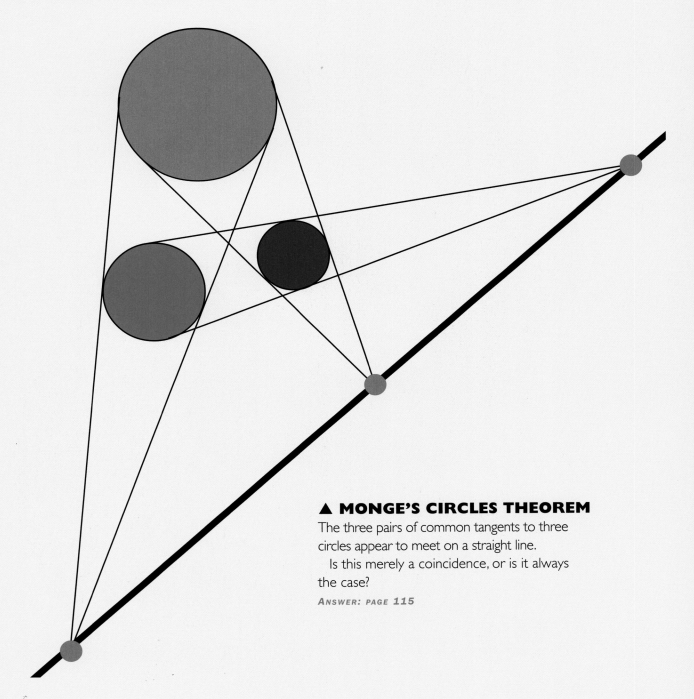

▲ MONGE'S CIRCLES THEOREM

The three pairs of common tangents to three circles appear to meet on a straight line.

Is this merely a coincidence, or is it always the case?

ANSWER: PAGE 115

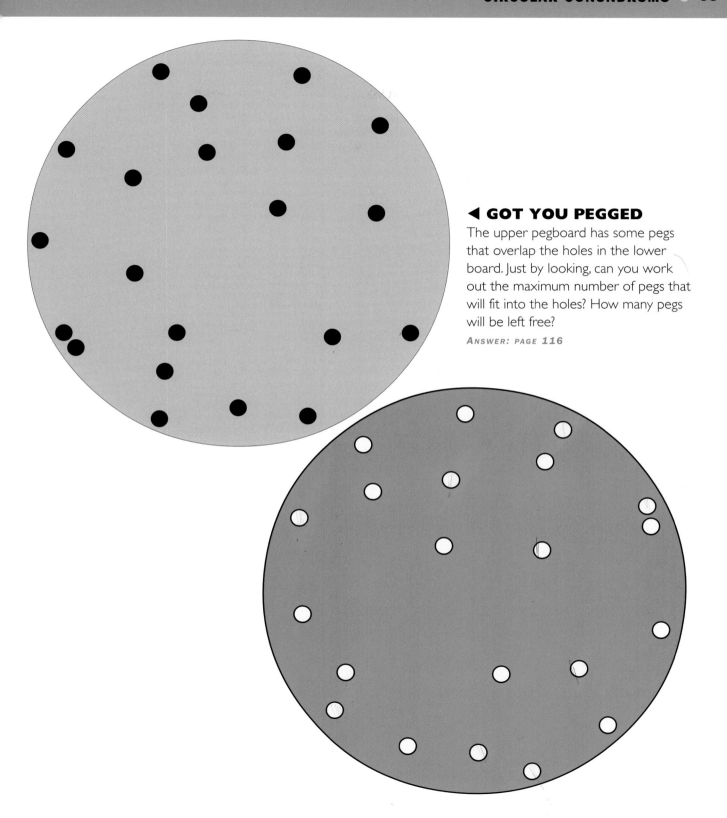

◄ GOT YOU PEGGED

The upper pegboard has some pegs that overlap the holes in the lower board. Just by looking, can you work out the maximum number of pegs that will fit into the holes? How many pegs will be left free?

Answer: page 116

Curves that have the same width in every direction are called curves of constant width. Any curve of constant width can turn between two fixed parallel lines or within a square. Although some curves of constant width, such as the circle, are smooth, others have corners; and while some are highly symmetrical, others are quite irregular. As a matter of fact, any regular polygon with an odd number of sides can be rounded up to create a curve of constant width.

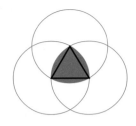

The Reuleaux triangle shown here has the smallest area for a given width of any curve of constant width.

◀ CONVEYANCING

These drinks are on an unusual conveyor belt. The cross-section of the rollers is shown here—one roller is circular, the others are more irregular. What will happen when the rollers start revolving?

ANSWER: PAGE **116**

✳ Rolling polygons and how to drill a square hole

Would it be possible to move heavy objects by means of rollers whose cross-sections are not circles, but some other sort of curves?

The answer is yes, and you can easily construct such a curve. First draw an equilateral triangle. Then, with each of the three corners as centers, draw the circular arcs passing through the other two corners. You obtain a triangular curve called the "Reuleaux triangle," named for the man who discovered it in 1875.

The width of a Reuleaux triangle in every direction is equal to the side of the equilateral triangle on which it was constructed.

Any curve of constant width can be revolved between two fixed parallel lines. In fact, it can be revolved within a square, each side of which is touched by the curve at every stage of its revolution. The side of the square is equal to the width of the curve. For these reasons, the curve also forms the basis of

a special patented drill that can drill a square hole.

A screw with a cross-section in the form of the Reuleaux triangle has been patented in the U.S. It is self-tapping and rolls its own thread in a wide variety of metals and plastics without making cuttings or chips.

The interesting properties of curves of constant width have been extensively investigated. One startling property, not easy to prove, is that the lengths (perimeters) of all curves with constant width have the same length. Since the circle is such a curve, the perimeter of any curve of constant width must of course be "π x d," the same as the circumference of a circle with the diameter "d." This is the astonishing theorem of the Russian-born mathematician Hermann Minkowski.

Many countries issue non-circular coins that are curves of constant width, such as the seven-sided 50p coin in Britain. It is vital that the coin has a constant width, otherwise it would be impos-

sible to measure its size accurately—an important feature in the design of vending machines.

Failure to recognize the existence of curves of constant width has had disastrous consequences in industry and has enormously increased the importance of inspecting machine components for true shape.

Take, for example, the hull of a submarine. Could it be tested for circularity just by measuring its maximum width in all directions? The answer is no, because such a hull can be very lopsided and still pass a test of this sort. It is for this reason that the circularity of a submarine hull is always tested by applying curved templates.

Michael Goldberg is the recognized leading expert on the subject. He has introduced the term "rotor" for any convex figure that can be rotated inside a polygon while at the same time touching every side.

◀ BUMPY RIDE IN CAT-ENA

I took my bike on holiday to the city of Catena in Sicily. I had a bumpy ride because all roads in Catena had equally spaced identical bumps of a certain shape and size as shown below.

I was very surprised to see that the Catenarians had no such problems. They were cycling happily on their bicycles, having perfectly smooth rides. Can you give an intelligent guess as to how they achieved this?

ANSWER: PAGE 116

▶ CURVES OF CONSTANT WIDTH

A circle revolves between two fixed parallel rails, as shown. A revolving ellipse, however, alters the distance between two parallel rails between which it revolves.

Which of the five red curves on the right turn like a circle and which like an ellipse?

ANSWER: PAGE 116

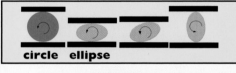

circle   ellipse

If you have a chain or necklace handy, why not try out the fun experiment below? You may learn something about the beautiful catenary curve.

▼ CATENARY:
The Gravity Curve

If you have a long thin chain, hold it at two points, overlapping it with the two points on the page and holding the page vertical, so you can see the freely hanging chain overlapping the curve on the page. In the 17th century, when Galileo did something similar he thought the curve he saw was a parabola. He was wrong. The curve is not a parabola—it is a catenary or "chain curve."

The word "catenary" comes from the Latin word **catena**, meaning "chain."

The catenary is the curve formed by a chain suspended freely between two fixed points. The mathematical expression used to describe this kind of curve is pretty complex and uses a form of trigonometry different from the sine, cosine, and tangent we learned at school.

The curve is produced by physical forces, those of gravity and tension, producing an esthetically pleasing result. The beautiful memorial known as the Gateway Arch, in St. Louis, Missouri, is an approximate "inverted" catenary, the ideal arch shape to support its own weight.

Every block in a catenary arch is held in place by the neighboring adjacent blocks. The forces of compression between every pair of blocks keeps the shape strong and firm. The blocks at the bottom are more vertical because they have more weight to support from the blocks above.

Like the parabola, all catenaries are of the same shape, differing only in size and how much of the curve is chosen. Pulling a chain or a rope taut does not change the shape of the curve, it merely gives you a piece from a larger catenary.

The catenary is characterized by its weight uniformly distributed along its length. What will happen to the shape of the curve if you tie and suspend two equal weights at equal horizontal distances apart?

ANSWER: PAGE **117**

PIET HEIN (1905–1996)

Piet Hein was a Danish all-around genius of wide-ranging interests. His depth of thinking is comparable to Leonardo da Vinci and Buckminster Fuller.

He is known for his thousands of short poems called "Grooks," containing his wisdom and philosophy, and for his activities as an underground resistance fighter during World War II.

He created the famous Soma cube, and games such as Hex, Morra, Nimbi, TacTix, and many others. The Soma cube and Hex are classics of topology.

Confronted with the traffic problem of Sergel's Square in Stockholm (see box at right), he perfected a new geometrical curve, the "Superellipse" and its three-dimensional counterpart, the "Superegg."

With his beautiful furniture designs he also gained the "Scandinavian design" international acclaim and acknowledgment.

❓ DID YOU KNOW?

When a Superellipse revolves about its long axis we get a surface of revolution— a Superegg, see right—an interesting 3-D solid that balances on both ends.

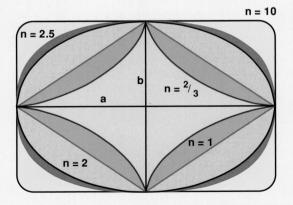

▶ CONIC SECTIONS

The ancient Greek scholar Apollonius, in his book **Conics** (dated 225 B.C.), revealed that a circle-based double cone can be cut to form a family of curved shapes by planes of suitable inclinations.

If you cut the three double cones by the four planes as illustrated (1–4) what will be the resulting curves?

ANSWER: PAGE 117

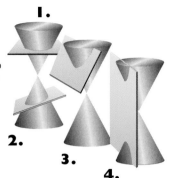

The superellipse of Piet Hein

You may know that the formula for graphing a unit circle is $x^2 + y^2 = 1$. You can change this into the formula for an ellipse (oval) shape by adding in two extra constant numbers, a and b:

$$(x/a)^2 + (y/b)^2 = 1.$$

The French mathematician Gabriel Lamé (1795–1870) wondered what would happen if we changed the powers from 2 to other values of "n": $(x/a)^n + (y/b)^n = 1$

If $n = 0$, we obtain a pair of crossed lines

If n is less than 1 ($n < 1$), we have a four-pointed star (asteroid)

If $n = 1$, we have a diamond

If $n = 2$, we have our ellipse again

If $n > 2$, the ellipse becomes more and more rectangular in nature.

For the specific case of $n = 2.5$, Piet Hein described the resulting shape as a "Superellipse" (see box), which he applied in many ways. For example, he used it to redesign the traffic flow around Sergel's Square in Stockholm as a compromise between the harsh edges of a rectangular layout and the wasted space of a circular ring-road. He also used a 3-D extension of the shape in his furniture design.

n = 10
n = 2.5
b
n = ²/₃
a
n = 2
n = 1

A parabola is defined as the path or locus of a point that moves so its distances from a fixed point (focus) and a fixed line (directrix) are equal. It is one of many curves found in nature.

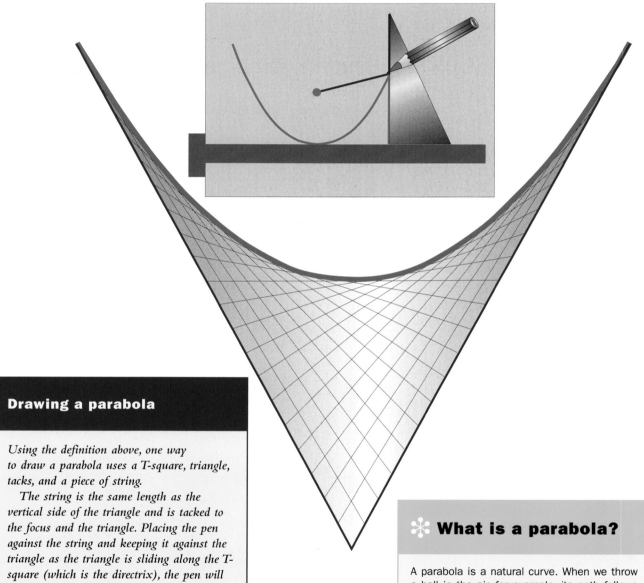

Drawing a parabola

Using the definition above, one way to draw a parabola uses a T-square, triangle, tacks, and a piece of string.

The string is the same length as the vertical side of the triangle and is tacked to the focus and the triangle. Placing the pen against the string and keeping it against the triangle as the triangle is sliding along the T-square (which is the directrix), the pen will draw a parabola as shown.

A parabola can be drawn as the envelope of two lines by connecting opposite points by straight lines.

❊ What is a parabola?

A parabola is a natural curve. When we throw a ball in the air, for example, its path follows a parabola. A spout of water takes the same shape. The shape has some interesting physical properties, making it ideal for reflectors in headlights and the dishes of radiotelescopes.

▼ PITCH UP THE PARABOLA

The suspended weights hanging below from this horizontal rod form a parabola, a curve that is one of the conic sections (see page 57).

What will be the shape of the curve formed by the suspended weights when the angle of the rod changes to 20 degrees from the horizontal?

ANSWER: PAGE 118

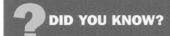

? DID YOU KNOW?

A parabola is the perfect shape for whispering galleries.

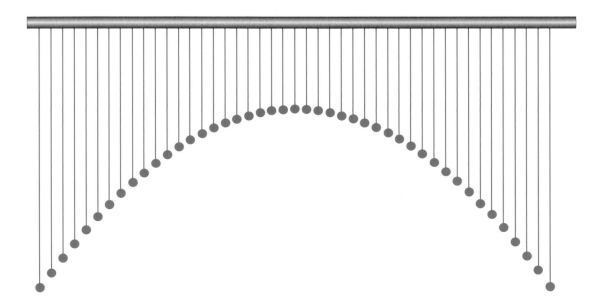

✳ Parabolas and puzzles

The parabola, a shape that joins up all the points which are the same distance away between a given point and line, is used in some puzzles. For example, if you were stranded at sea but knew you were equidistant from a lighthouse (at the focus) and the shore (the directrix), the parabolic line shows your rescuers all the possible locations at which you could be.

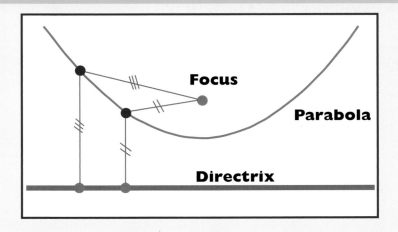

Focus

Parabola

Directrix

Some primitive species had only three numbers in their vocabulary: "one," "two," and "many." Would that have affected their ability to do these puzzles?

▶ HOW MANY HEXAGONS?

How many hexagons can you count in this image?

ANSWER: PAGE 118

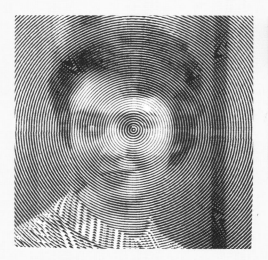

Computer manipulation

This picture shows computer treatment of a photograph.

The brightness was transposed into the line thickness of the concentric circles. (Manfred Schroeder)

◄ GOING ROUND IN CIRCLES?

How many circles do you think make up this intricate engraving?

(Engraving by Claude Mellan, France, 1649)

ANSWER: PAGE *118*

Many years ago I invented several drawing machines of which the Kinemat was the simplest, and which you can easily build by following the blueprints and descriptions provided here.

❋ KINEMAT—Art Drawing Machine

Though seemingly simple, this ingenious device can be used to draw a practically endless number of esthetic families of complex curves. Mathematically, these curves have been named "tricircular-sextrics," which means they are described by equations containing terms up to x^6.

The secret to obtaining limitless pleasing configurations lies in creating systematic rules and procedures, leading to often surprising outcomes.

The shapes of the resulting families of curves and their location on the page depend upon the proportions of the linkages and the position at which the puzzler chooses to place the pen from the endless possibilities available. When these parameters are changed slightly, a completely new and different family of curves is the result.

Operation: The parts of the Kinemat can be found on pages 64–65. The coupler disk has a variety of points at which to place the pen. Both the crank and rocker arms have four holes and their pins can be placed in any two of the holes of the base plate. The two arms can be placed at different lengths.

The coupler disc can now be mounted on the two fixed pins of the two arms.

A pen can be placed in any of the holes of the coupler disc. By manipulating or "cranking" one or both pins of the arms, the coupler disk with the pen will be forced to move and produce the curves, while the arms will rotate in a circle or oscillate in circular arcs.

Sometimes it may be more convenient to "drive" the pen itself at the selected coupler point, or the disk itself. Drawing the coupler curves at the beginning requires a bit of skill and practice, but, with time, the results are very rewarding. Creating drawings and discovering an infinite number of new forms and patterns soon becomes addictive.

The curves produced from using the Kinemat are inherently esthetic and quite complex. They contain no true circular arcs, nor true straight lines. They have as many shapes as points selected; they may have sharp curves, easy curves, crossovers, double points, near-straight sections, cusps, etc., as the examples show.

The 12 drawings shown opposite were all created by children, and some of the drawings were later colored in by them.

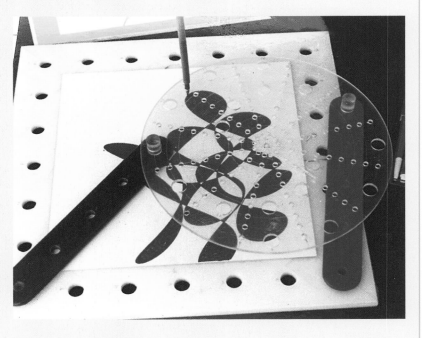

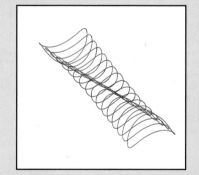
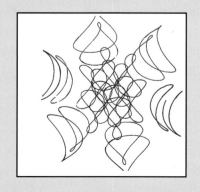
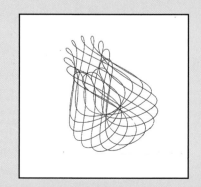
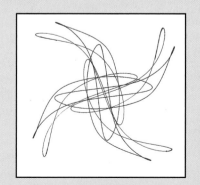

▼ Kinemat drawing machine

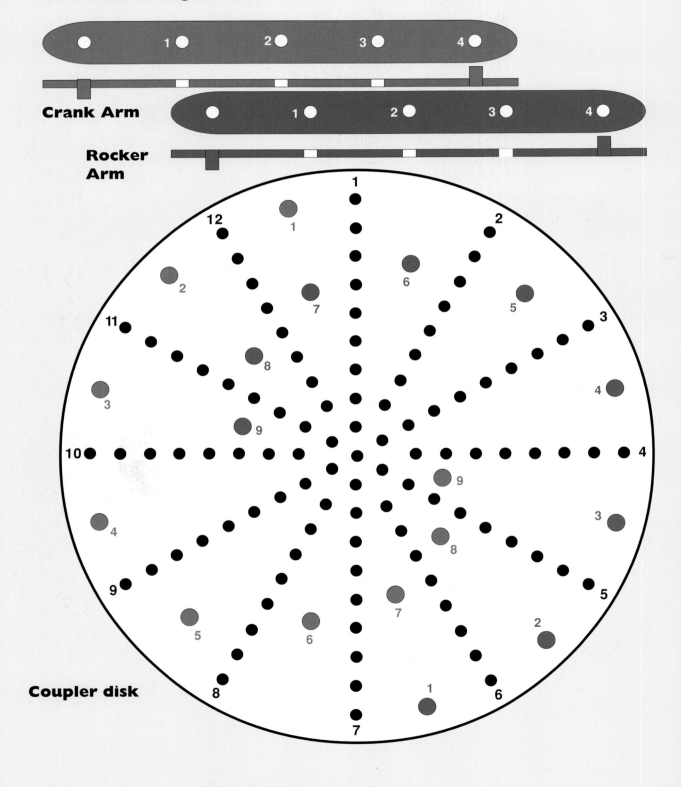

Crank Arm

Rocker Arm

Coupler disk

Kinemat drawing board

An amusing mathematical proof called the Pancake theorem states that if you lie two pancakes side-by-side, there always exists one straight line that will cut both pancakes into equal halves. Amazingly, the pancakes do not even have to be circular. Luckily for you, the properties of a circle will help you here.

▼ PUZZLE WITH A TWIST

This emergency escape staircase in the helipad structure is in the form of a helix spiral, traveling exactly four times around the building as shown.

The circumference of the round building measures 40 units and its height is 120 units.

Can you figure out the length of the spiral staircase?

ANSWER: PAGE **118**

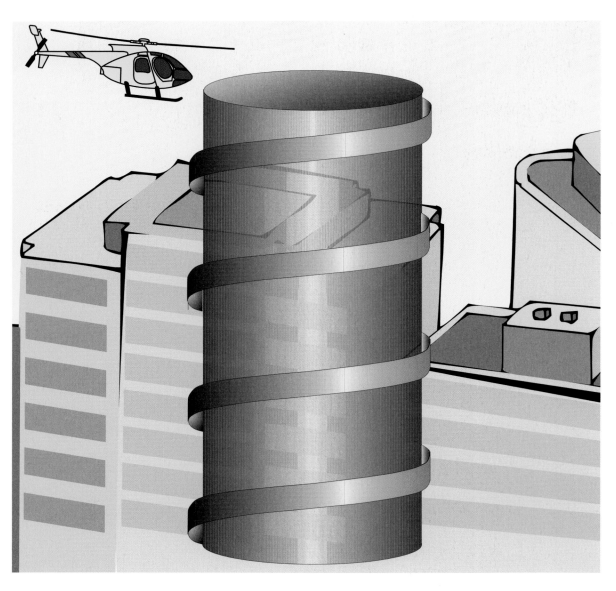

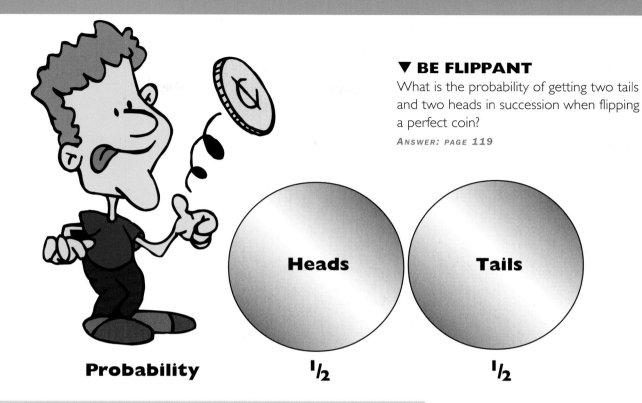

▼ BE FLIPPANT

What is the probability of getting two tails and two heads in succession when flipping a perfect coin?

ANSWER: PAGE 119

Probability

Heads ¹/₂

Tails ¹/₂

◄ ELUSIVE ELLIPSE?

How can this man, sitting near his table, create an ellipse without touching his pen, ruler, compass, or computer?

ANSWER: PAGE 119

These two puzzles have the same kind of theme but the approach to each is very different. One is very logical and calculating; the other requires a more visual and instinctive approach.

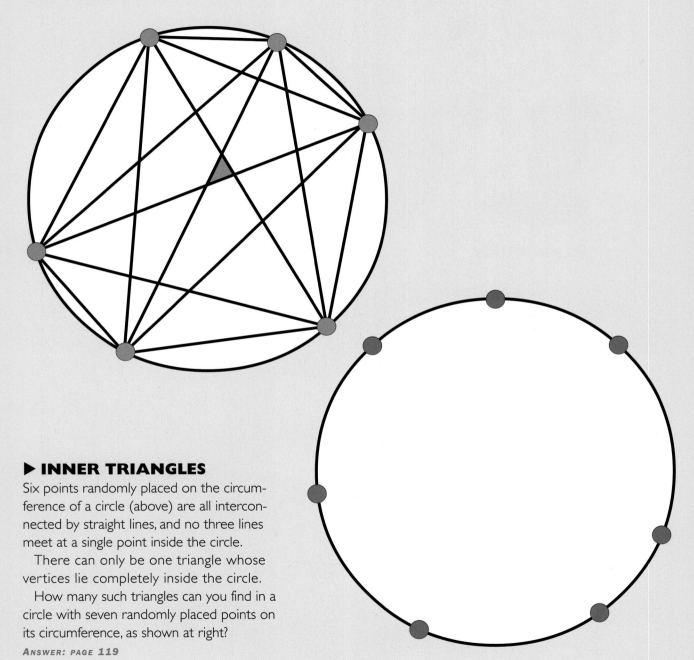

▶ INNER TRIANGLES

Six points randomly placed on the circumference of a circle (above) are all interconnected by straight lines, and no three lines meet at a single point inside the circle.

There can only be one triangle whose vertices lie completely inside the circle.

How many such triangles can you find in a circle with seven randomly placed points on its circumference, as shown at right?

ANSWER: PAGE 119

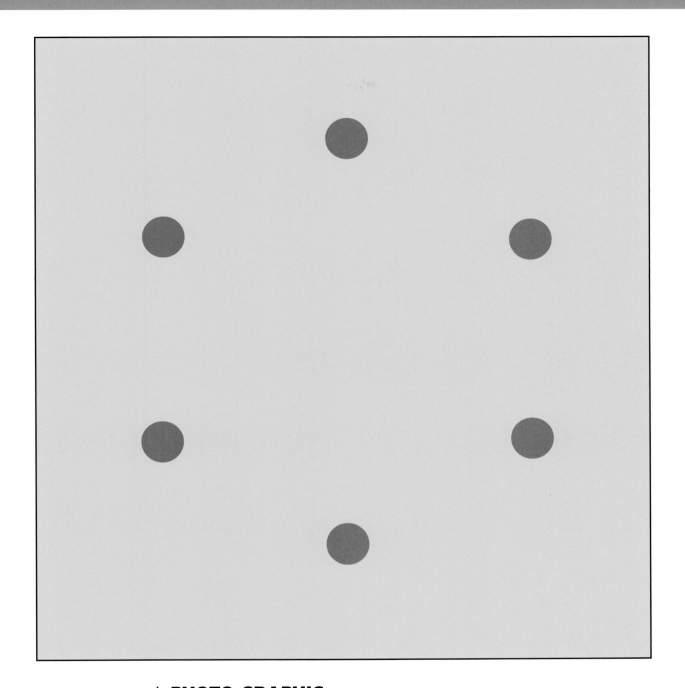

▲ PHOTO-GRAPHIC

Out of a total of 15 possible connecting lines between the six points above, what is the maximum number of edges (interconnections between two points) you can draw without crossings (intersections)? Hint: The lines you draw do not have to be straight.

ANSWER: PAGE 119

nsects and arachnids are all over the place here. Good job we've kept them apart, otherwise the insects would all be eaten and there wouldn't be much of a puzzle left!

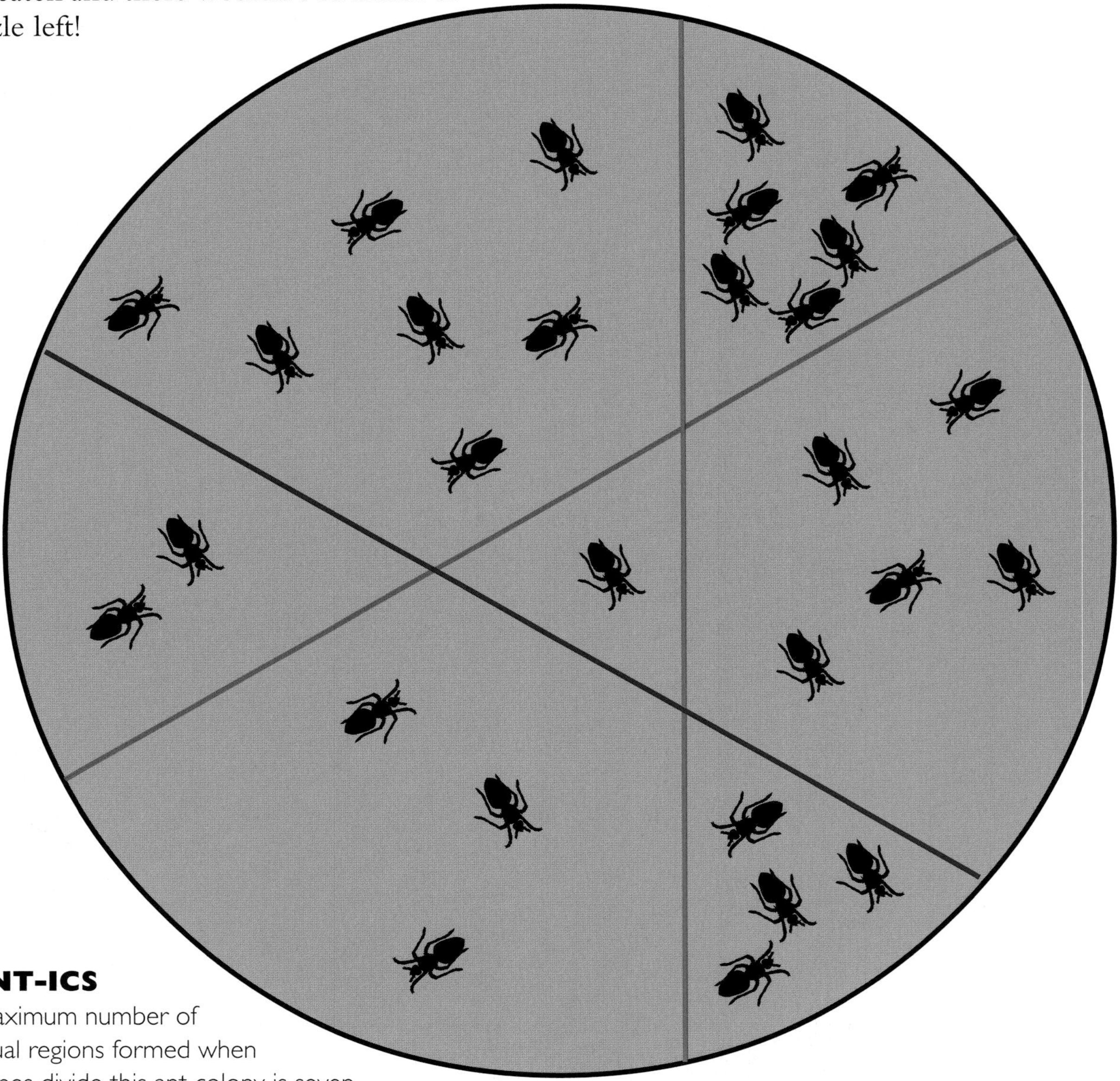

▶ ANT-ICS

The maximum number of individual regions formed when three lines divide this ant colony is seven as shown. Between one and seven ants occupy each of the seven regions.

Can you place the seven groups of ants in the seven regions so that for each of the lines, the total number of ants on either side are all the same?

How many solutions can you find?

ANSWER: PAGE 120

◀BUTTERFLY COLLEC-TION

Using five straight lines, can you separate the 16 butterflies into 16 separate compartments, with each compartment containing one single butterfly? In the trial below, five lines were drawn, but only six butterflies could be separated, while the rest are occupying compartments containing more than one each.

ANSWER: PAGE 120

◀ WEB WEAVERS

From each of the six spiders on the periphery of the web, draw a straight line through the web, so as to surround and enclose each insect in a solitary entrapment.

When the six lines are drawn, each of the 18 insects must be alone in a closed compartment created by the six intersecting lines.

ANSWER: PAGE 120

You may remember from school that lines can be generalized into the form y = mx + c (m = slope, c = the place where the line crosses the y–axis). Below we take a look at more interesting forms of graphical equations.

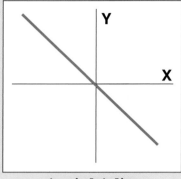

straight line

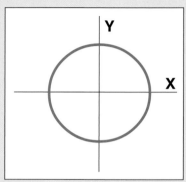

circle

◀ **CURIOUS CURVES**

Lines on a graph join up all those points (x, y) that form a valid equality when those combinations of x and y are placed into the given equation.

 Can you identify the curves corresponding to the six equations below?

ANSWER: PAGE 121

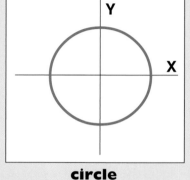

ellipse

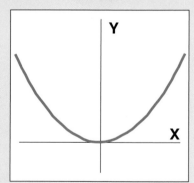

parabola

$$x^2 + 4y^2 = 4$$

$$(x^2 + y^2)^3 = 4a^2 x^2 y^2$$

$$x + y = 0$$

$$x^2 = 4y$$

$$x^2 + y^2 = 4$$

$$x^3 + y^3 = 3axy$$

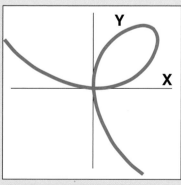

folium of Descartes

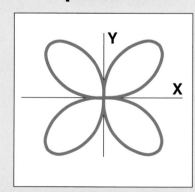

rose of Grandi

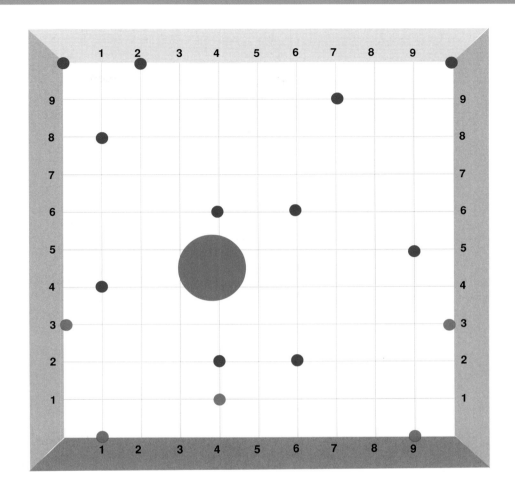

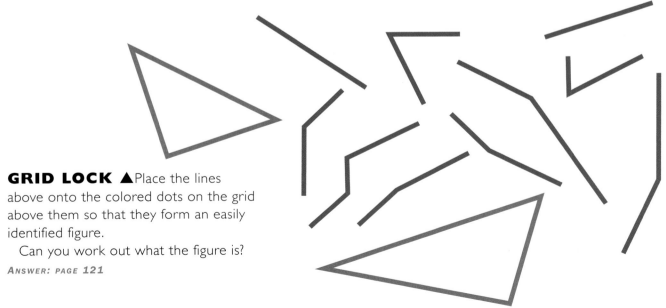

GRID LOCK ▲ Place the lines above onto the colored dots on the grid above them so that they form an easily identified figure.

Can you work out what the figure is?

ANSWER: PAGE 121

It seems a shame that so many puzzle books are in color and yet the puzzles themselves rarely use color as part of their principle. Luckily, this isn't one of those books!

▼⊠ MAGIC COLOR SQUARES QUARTET

Place the 16 squares on the 2-by-2 gameboard to create four Latin color squares, in which four different colors form each horizontal and vertical row.

ANSWER: PAGE 122

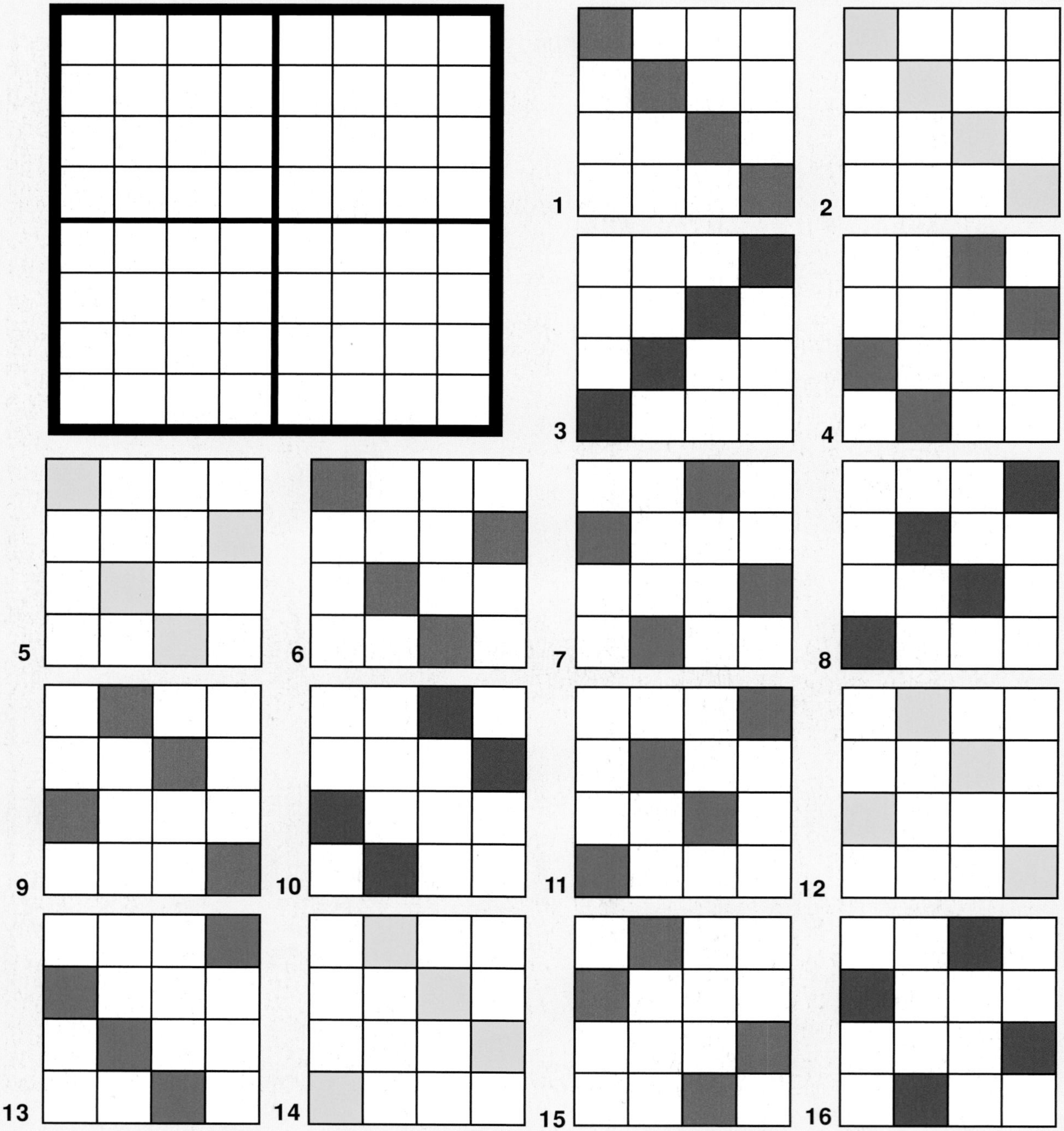

1

2

3

4

5

6

7

8

9

10

11

12

13

14

15

16

▼ RED–GREEN–BLUE

The object of this puzzle is to create the shortest path from start to finish following the pattern red–green–blue, red–green–blue all the way.

It is permissible to revisit points, but retracing the colored lines is not permitted.

Can you also solve the puzzle as green–red–blue?

ANSWER: PAGE 122

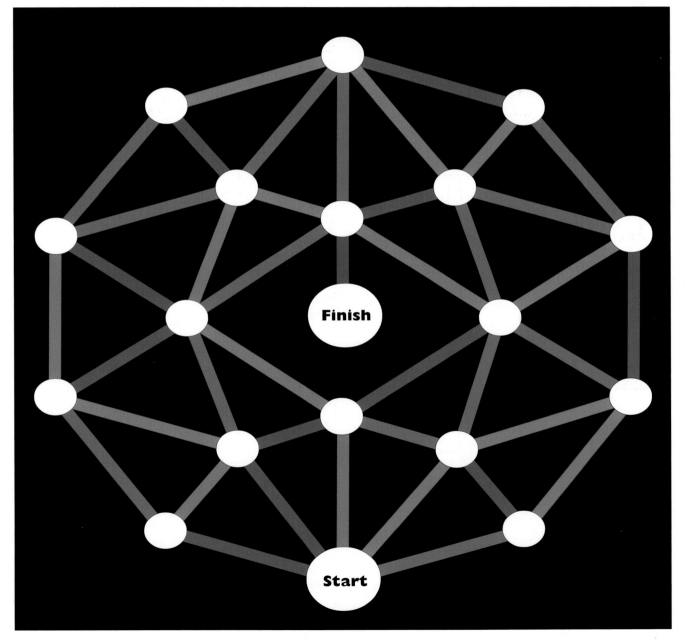

This is a paper-and-pencil game for two players. The object is for either of the players to cross the board in a continuous route, each time connecting two grid points with a straight line, using a move like that of a knight on a chessboard.

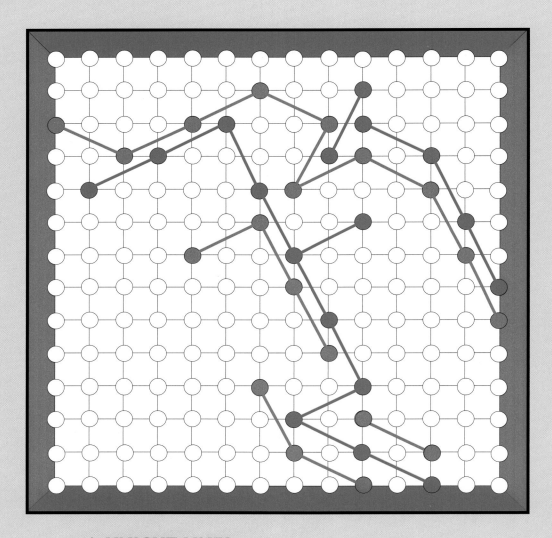

▲ KNIGHT LINES

The game can be played on a square grid, using two colored markers as shown.

Player 1 has to connect the red sides of the gameboard, while player 2 has to connect the green sides.

The routes are not permitted to cross. The sample game above shows the red player winning after a tough fight.

▼ THE GAMEBOARD

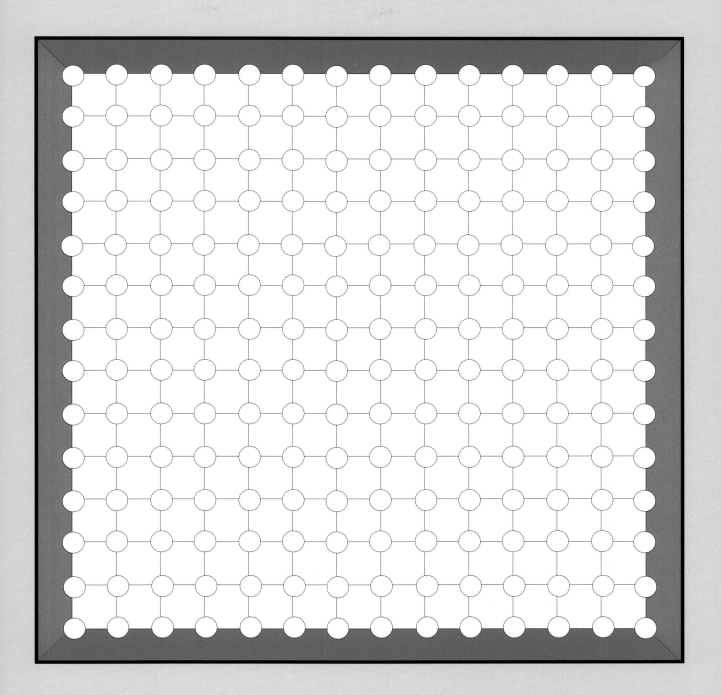

We're traveling all the time. During a racket sports game, a player will run over a mile. And the Earth is traveling through the universe at a tremendous pace. So, don't get dizzy when you find the right routes in these puzzles.

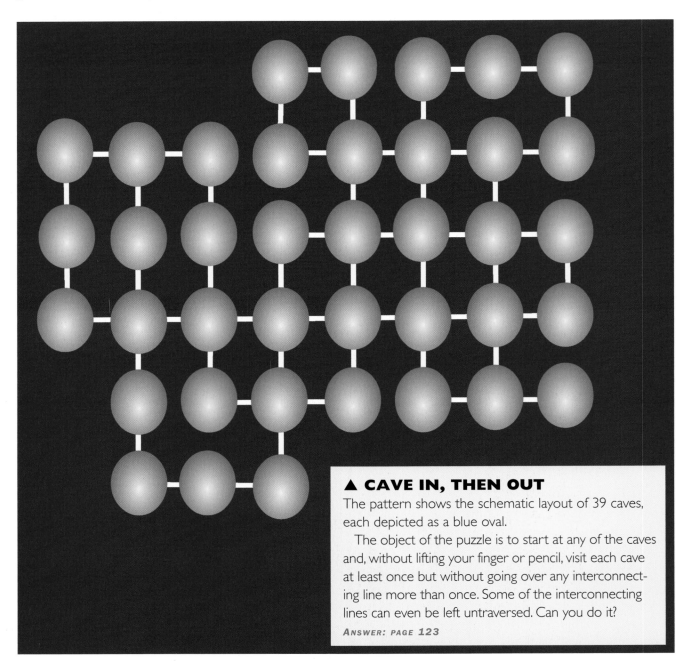

▲ CAVE IN, THEN OUT

The pattern shows the schematic layout of 39 caves, each depicted as a blue oval.

The object of the puzzle is to start at any of the caves and, without lifting your finger or pencil, visit each cave at least once but without going over any interconnecting line more than once. Some of the interconnecting lines can even be left untraversed. Can you do it?

ANSWER: PAGE 123

▼ RECONNAISSANCE

The high-altitude reconnaissance plane crossed the red boundary 12 times before returning to its home base. The next time it wanted to make a similar mission, crossing the boundary only 11 times before returning. Can you suggest a path for such a mission?

ANSWER: PAGE 123

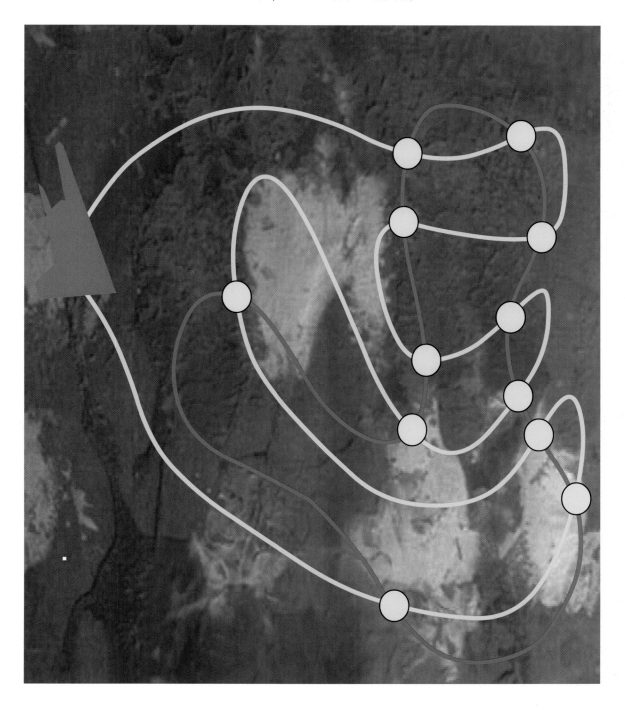

Marilyn Monroe's portrait is one of the most widely duplicated snapshots of all time. So much so, in fact, that you can reduce the image to its bare essentials and still recognize it.

▶ A DOTTY ACTRESS

Can you guess how many dots were needed to get a good likeness of Marilyn? Try making a rough guess then see if your answer falls within 25 dots either way.

ANSWER: PAGE 123

▶ ALPHABET SOUP

Can you tell the difference between the red and blue letters in each of the three groups of capital letters of this alphabet?

Two letters are missing from each group and they are outside the chalkboard as shown. Can you place these letters back into the groups in which they belong?

ANSWER: PAGE 124

Group 1

Group 2

Group 3

▶ ANY WHICH WAY

Can you visualize and draw an object that could hermetically fit and close either of the three holes in this children's cube toy?

ANSWER: PAGE 124

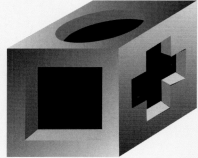

Construction toys such as Erector sets are still popular today. You may need to resort to them to solve these puzzles about lines and linkages.

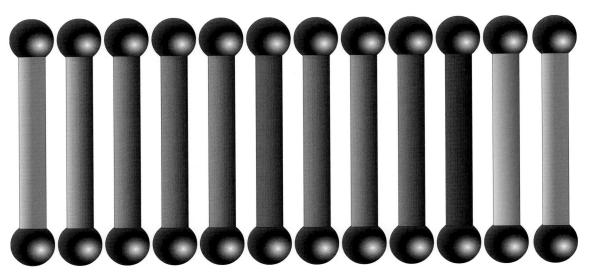

▶ SKELETON CUBES

Many problems of combinations and permutations are of great interest in recreational mathematics because of the large number of different ways in which combinations of even a few obects can be arranged.

For this reason, an ingenious magnetic construction toy has proved very popular. Different-colored rods can be magnetically attached to each other in any desired way.

Twelve different-colored rods of equal length can be magnetically attached to form a skeleton cube as shown.

Can you guess how many different-colored cubes you could form from the 12 magnetic rods?

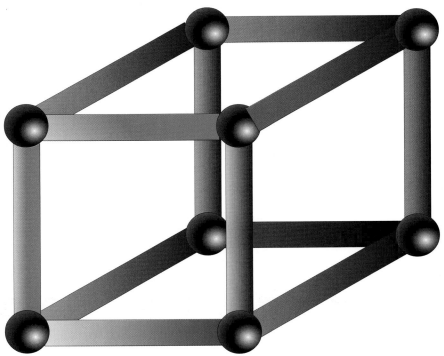

ANSWER: PAGE **125**

▶ PANTOGRAPH

Can you imagine the appearance and guess the size of the path described by the red pencil point when you trace a green line with the tracing point?

ANSWER: PAGE 125

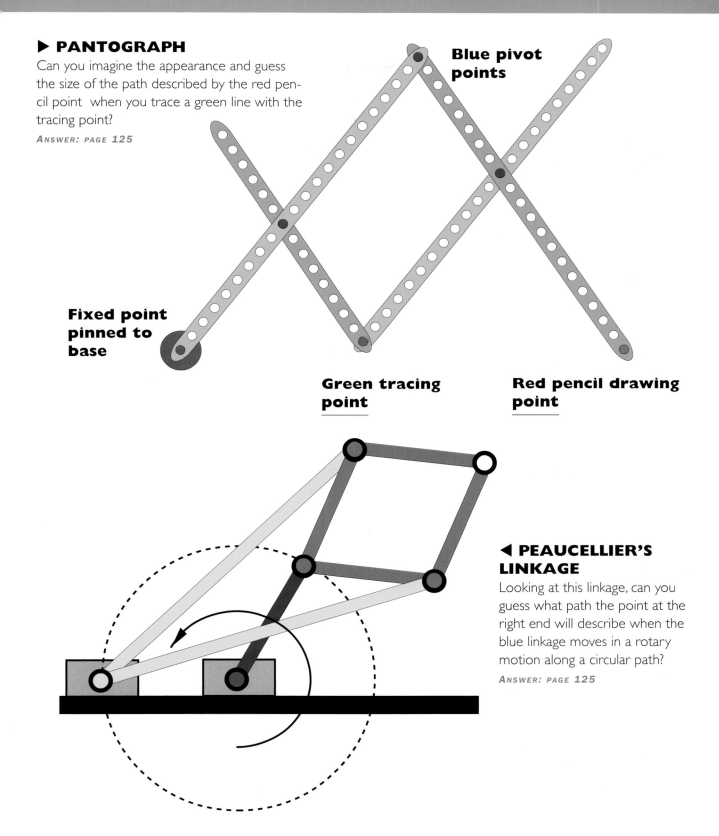

Blue pivot points

Fixed point pinned to base

Green tracing point

Red pencil drawing point

◀ PEAUCELLIER'S LINKAGE

Looking at this linkage, can you guess what path the point at the right end will describe when the blue linkage moves in a rotary motion along a circular path?

ANSWER: PAGE 125

Start

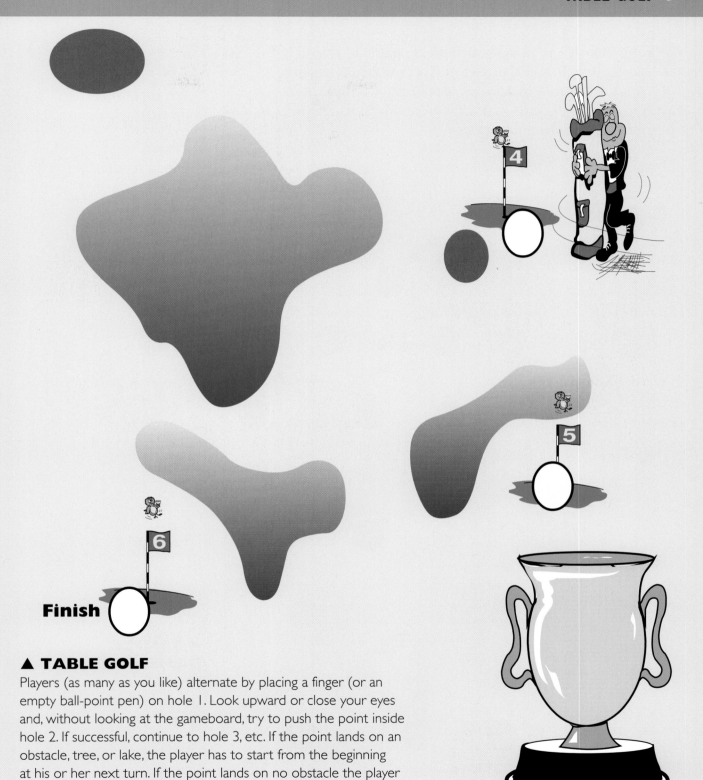

Finish

▲ TABLE GOLF

Players (as many as you like) alternate by placing a finger (or an empty ball-point pen) on hole 1. Look upward or close your eyes and, without looking at the gameboard, try to push the point inside hole 2. If successful, continue to hole 3, etc. If the point lands on an obstacle, tree, or lake, the player has to start from the beginning at his or her next turn. If the point lands on no obstacle the player continues from that position next time and has another attempt at placing the ball "into" the hole.

Pong Hau K'i is an ancient Chinese game a little like tic-tac-toe. It is also played in Korea under the name "Ou-moul-ko-no." It appears deceptively simple, but even such simple games can offer a considerable amount of subtleties and need for analysis.

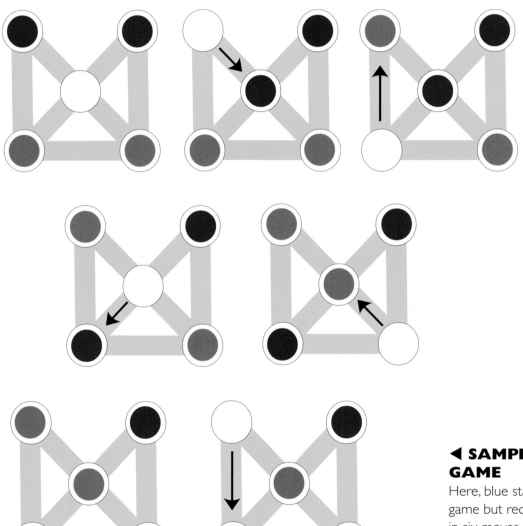

◄ SAMPLE GAME

Here, blue starts the game but red wins in six moves.

▼ THE GAMEBOARD

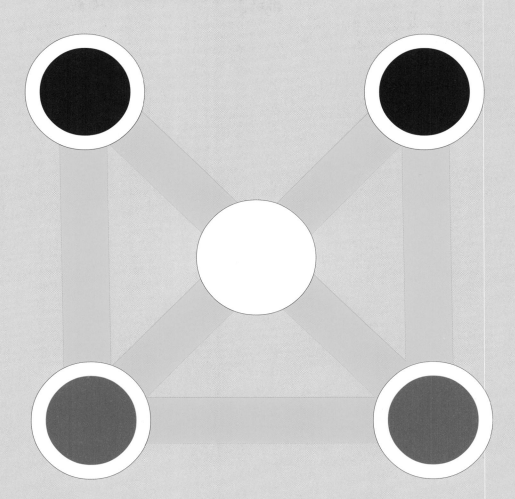

▲ PONG HAU K'I

Each player has two playing pieces that are initially placed on the gameboard as shown. Players take turns moving one of their pieces along a line to an adjacent empty circle. Jumping over another piece is not allowed. The object of the game is to trap the opponent's pieces.

1. Is one set of pieces better than the other?
2. How many possible positions are there in a game?
3. What are the best strategies for the players?
4. Can you force a draw in a game?

Play the game with your friends to answer these questions.

Astrophysicists say that the universe possesses four dimensions—three of space and one of time—and some recent theories have suggested that there may be even more dimensions exerting an influence at the subatomic scale. How can we begin to understand hypothetical higher dimensions? By getting outside our normal system—in this case, trying to imagine a world that has only two dimensions.

✳ Flatland—a world of two dimensions

In 1884, Edwin A. Abbott, an English clergyman and popularizer of science, made a beautiful attempt at describing a world made up of only two dimensions. In his satirical novel, called *Flatland*, the characters are basic geometrical figures gliding over the surface of an infinite two-dimensional plane—a vast tabletop. Apart from negligible thickness, Flatlanders have no perception of the third or any higher dimension.

Although Abbott himself did not describe any of the physical laws or technological innovations of a world such as Flatland, his book spawned sequels that tackle just those issues. One such book, *An Episode of Flatland*, written by Charles Howard Hinton in 1885, cleverly extends Abbott's original idea.

The action in Hinton's book takes place on the apparently two-dimensional planet, Astria. Astria is simply a giant circle, and its inhabitants live on the circumference, forever facing in one direction. All males face east, and all females face west.

To see what is behind him or her, an Astrian must bend over backward, stand on his or her head, or use a mirror.

Astria is divided between two nations, the civilized Unaeans in the east and the barbaric Scythians in the west. When the two nations go to war the Scythians have an enormous advantage: They can strike the Unaeans from the back. The unfortunate and helpless Unaeans are driven to a narrow region bordering the great ocean. Facing complete extinction, the Unaeans are in fact saved by a scientific advance: Their astronomers have discovered that their planet is round. A group of Unaeans cross the ocean and carry out a surprise attack on the Scythians, who have never been attacked from the rear. The Unaeans are thus able to defeat their foes.

Houses in Astria can have only one opening. A tube or pipe is impossible. Ropes cannot be knotted, although levers, hooks, and pendulums can be used.

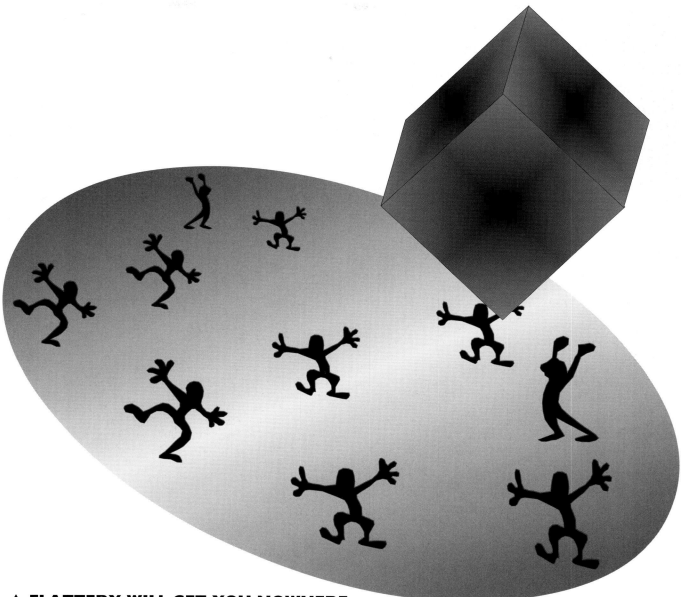

▲ FLATTERY WILL GET YOU NOWHERE

Imagine intelligent two-dimensional aliens confined to a two-dimensional surface world called Flatland. They are confined to Flatland not only physically but also sensorially—they have no faculties to sense anything out of their surface world.

In an event that occurs every 100,000 years, a three-dimensional cube collides with and passes through Flatland.

Can you describe how the unfortunate Flatlanders might observe this astronomical catastrophe?

ANSWER: PAGE 126

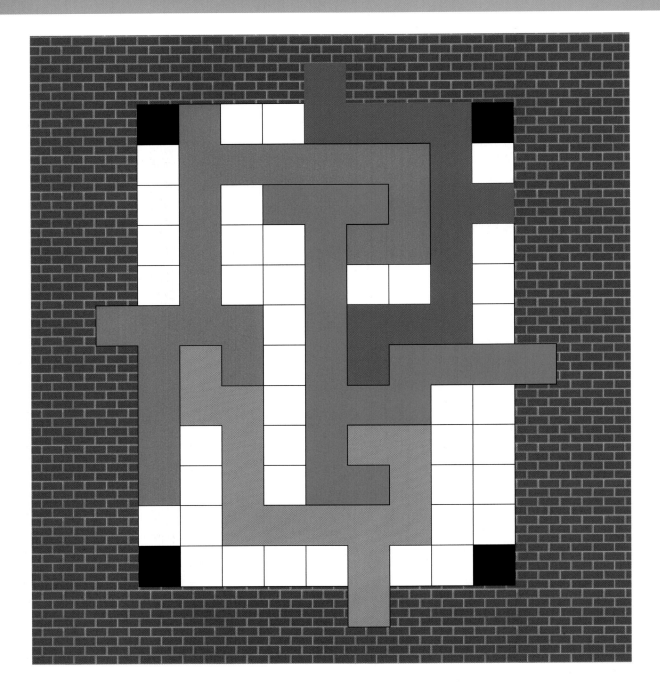

▲ **BANK RAID**

In the two-dimensional space of Flatland, the bank vault is an ingenious mechanism consisting of four sliding linkages, which can move up, down, left, and right only.

The four black squares are fixed at the four corners of the rectangular vault and the sliding linkages cannot cross them.

The object is to remove in succession the four linkages, one on each side of the vault to uncover it completely.

Which piece must be moved first? How many moves will it take you to open the vault?

ANSWER: PAGE 127

▼ ON THE RIGHT TRACKS

Railway tracks in Flatland are monorails. Eleven straight parallel railway tracks run through two cities. The tracks can connect the two cities without any intersections, which is advantageous for scheduling purposes. The leaders of a third city, not in line with the existing rail lines, have asked for some tracks to be relaid so the city can be connected to the others.

The tracks will be laid out so that one set is parallel in one direction, another is parallel in another direction, and a third set is parallel in a third direction. For efficient communication to exist, at least two tracks should run in any of the three directions. Note that railway lines cannot just stop on reaching the edge of a city.

Can you work our how to design the rail system in such a way that you create the fewest number of intersections?

ANSWER: PAGE 127

CITY C
Railway Station
Under construction

CITY A
Railway Station

CITY B
Railway Station

The cat's cradle string game is played by children around the world, including Inuit (Eskimo), African, and Australian kids, as well as Americans and Europeans. The puzzle opposite suggests a simpler version.

▼ ARE YOU COORDINATED?

A point on the two-dimensional plane of the paper can be determined as the intersection of two lines (coordinates).

Discover the picture produced by interconnecting in succession the 16 points obtained by plotting the coordinates.

ANSWER: PAGE 127

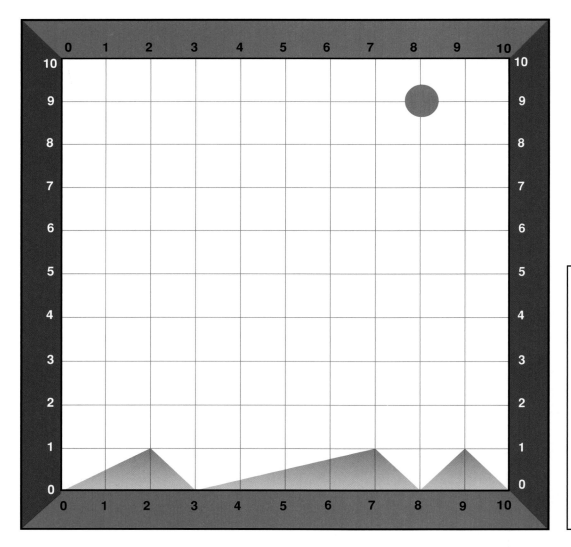

1	5	0
2	7	3
3	5	2
4	0	3
5	1	7
6	3	4
7	6	6
8	7	9
9	8	10
10	10	10
11	8	8
12	9	4
13	10	2
14	8	3
15	8	2
16	5	0

▶ CAT'S CRADLE

Imagine you have a piece of closed thin string forming a square with four fixed beads at the corners of the square, of unit length sides.

How many geometrically defined structures can you create from this flexible structure?

One of the structures is shown—holding opposite sides straight and parallel produces a family of rhombi, from the square opposite to a straight line of 2 units length, shown at the bottom of the page.

Can you find more structures that can be formed?

ANSWER: PAGE 128

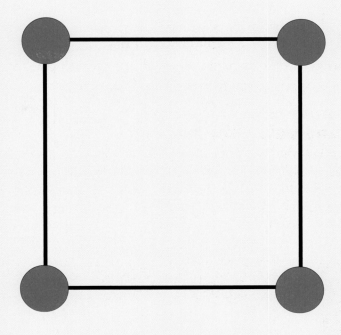

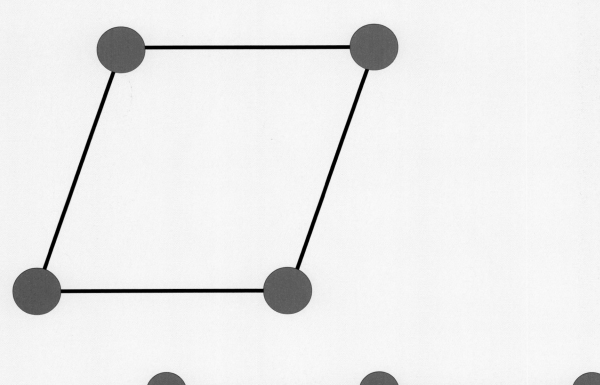

Subway maps are iconic pieces of design. The world's first underground railway in London now has a map based on a design by an electrical engineer called Harry Beck. See if you can make the right connections here!

▼ TIGHT SQUEEZE

Each player has two playing pieces; one player plays red and the other blue. Their starting positions are marked on the gameboard.

Players take turns moving one of their pieces along a straight line to an empty adjacent space. No jumping is allowed. A player loses the game when one of his or her colors becomes squeezed between the opponent's two colors: That is, when all three playing pieces are in a straight line, the loser's piece is in the middle, and no empty spaces are between them.

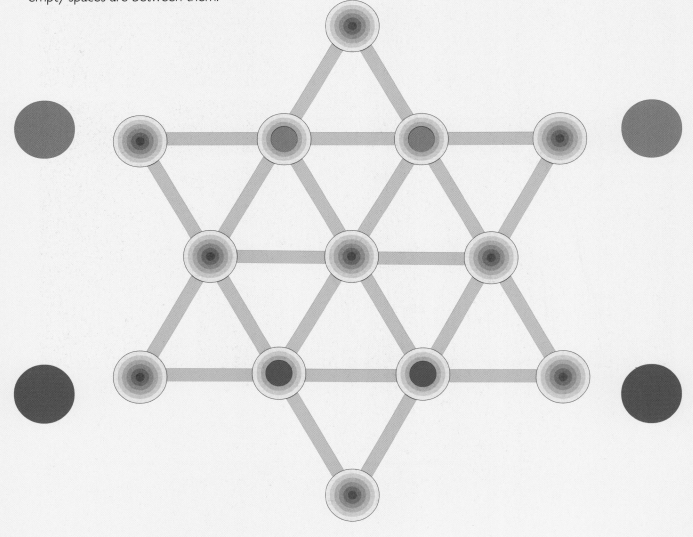

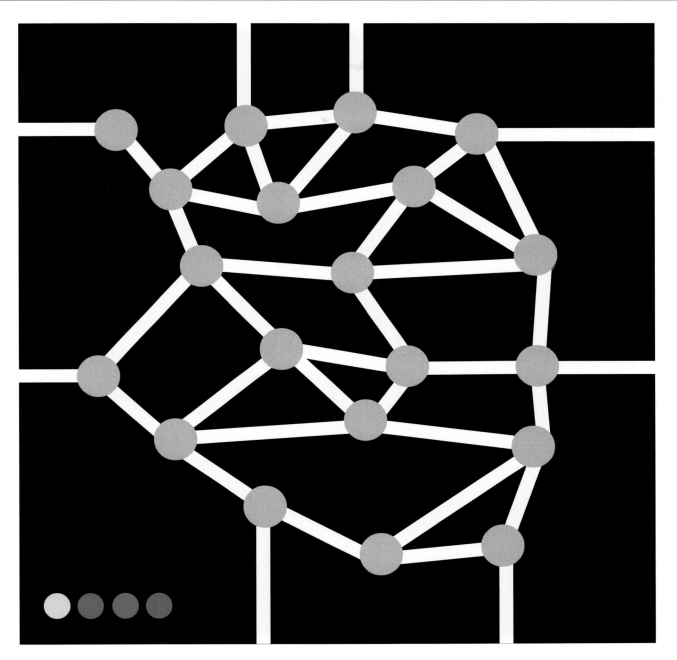

▲ THE PATH TO THE FINISH

You can complete this challenge alone or play it as a game.

The puzzle Using four colors, color the lines between the intersection points so that all lines meeting at these points are different.

The game Players alternate by coloring a line between two points. A player unable to color a line on his or her turn is the loser.

ANSWER: PAGE 128

OK, so it's the end of the puzzles, but try to put a brave face on it. Or even 13 brave faces on it.

▲ MOUTH TO MOUTH

The puzzle above could also be described as the game of vanishing faces.

The puzzle A total of 36 square tiles form a 6-by-6 square gameboard as shown. In the given configuration you can count 13 complete faces—nine smiling and four sad. Can you rearrange the tiles in the same configuration to make a face vanish—and have five smiling faces and seven sad?

How many other variations on the number of smiling and sad faces can you find?

ANSWER: PAGE 128

▼THE GAME

Using the 36 square tiles opposite, you can also play a game for two players. The object is to be the first to create a meandering red line connecting two opposite sides of the gameboard. In the game below, the bottom side of the gameboard is connected to the top.

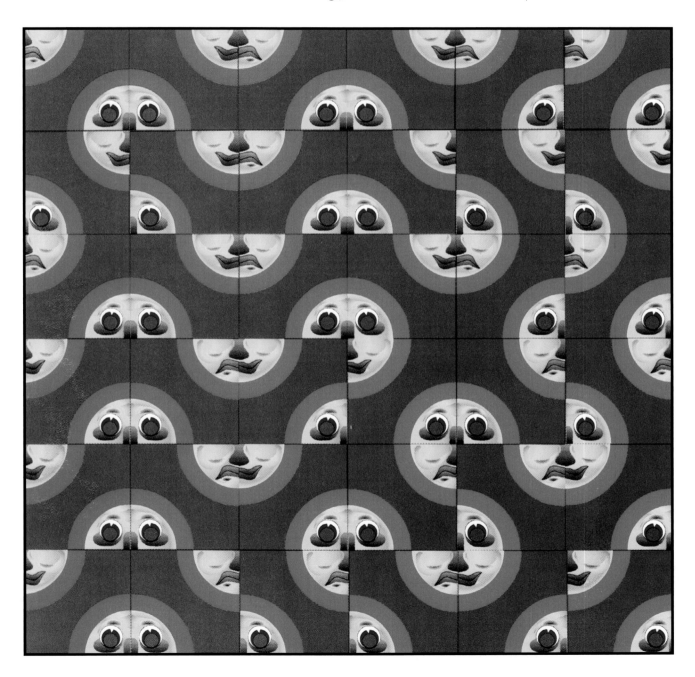

ANALEMMA (page 6)
Curve in the Sky

Sundial owners are familiar with the figure eight curve called the analemma. It provides them with a correction to solar time to allow for the changing rate at which the sun traverses the sky.

The loops are caused because the Earth's equator is approximately 23 degrees different from the plane in which the Earth orbits. In short, it has a wobble. The loops are of unequal size because the Earth's orbit is elliptical. It speeds up when it is turning the corner around the sun, and slows down at its farthest point from the sun.

In a model world where the Earth orbits the sun in a perfect circle, its equator on the same plane as its orbit, di Cicco's photograph would have been reduced to a single saturated dot.

The three streaks show the rising sun near the solstices and equinoxes: winter solstice and the shortest day (bottom right); the summer solstice and longest day (top left); the curve crosses on the celestial equator where we encounter the vernal and autumnal equinoxes.

A good picture is certainly worth a thousand words!

CONCENTRIC STARS (page 7)

If you have a camera you can easily create photographs like this. Fix your camera so that it is pointing at the Pole Star. Set the camera to "time exposure," leaving the shutter open for a long period of time. Stars will revolve around the Pole Star approximately once every day.

For an observer standing at a fixed position on Earth, the rotation of the Earth makes it appear as if the sky is revolving around it.

The Earth rotates about an imaginary line that passes through the North and South poles of the planet, and points in space to the Pole Star.

Aristarchus of Samos, an early Greek astronomer (about 310–230 B.C.), was the first person to suggest that the Earth revolved around the sun, rather than the other way around.

At the equator, the Earth is revolving at a speed of 1,034 miles (1,665 km) per hour.

▼ COMPLEX PLANETARY CURVES (page 7)

The early Greek mathematician Aristarchus suggested that the problem could be understood if all the planets, including the Earth, were to orbit the sun. But, after inciting much controversy, his argument was lost. (We only know of it today because Archimedes refers to it—disparagingly.)

During the 2nd century A.D. the astronomer Ptolemy summarized all the geographical knowledge amassed in his *Geographia*.

The Ptolemaic theory, putting the Earth at the center of the universe, went unchallenged for over 1,000 years.

Today, however, we know that the planets in the Solar system orbit the sun, and in elliptical orbits rather than circular paths.

Below is a model of successive views of an outer planet as seen from an inner planet as they revolve in concentric orbits about a sun, demonstrating the seemingly retrograde motion of the outer planet

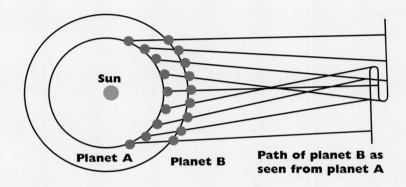

Sun

Planet A   **Planet B**   **Path of planet B as seen from planet A**

◀ STRING ALONG 2 (page 9)

This is the "1:3" web, although it has a grander name: the nephroid (kidney) curve.

▶ STRING ALONG 3 (page 11)

The solutions obtained by following rules 2 and 3 are shown on the right.

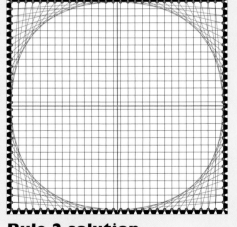

Rule 2 solution

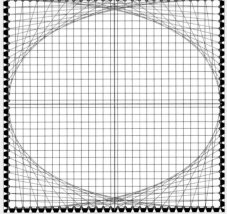

Rule 3 solution

▶ **BROKEN NECKLACE**
(page 12)

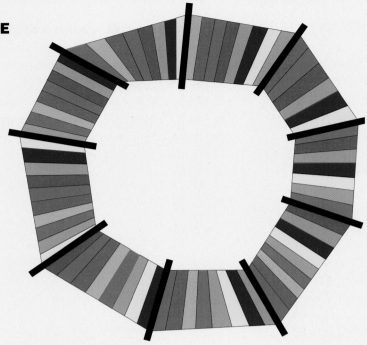

▼ **DNA (page 13)**

The two closed rings are mirror images of each other. The secret of joining the pieces is the color sequence shown in the top piece of both loops. This sequence is followed in the chain clockwise (right loop) and counterclockwise (left loop), with each piece starting with the next color of the sequence. (So the second piece starts with an orange segment, the third piece with a red segment, and so on.)

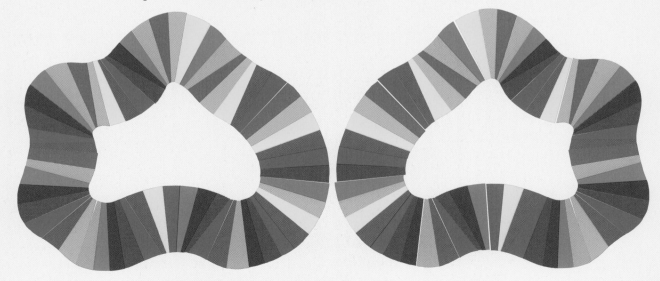

TO INFINITY ... THEN STOP (page 14)

The curvature of the Earth's surface is so negligible that it appears flat. The limit of decrease in curvature is a straight line. An infinite circle is therefore a straight line, believe it or not.

Something to think about: If a straight line is an infinite circle then where is its center?

HYPNOTIC GAZE (page 15)

If the center circle has a radius of one unit, the area will be equal to π units^2. The radiuses of the remaining circles have been chosen so that their areas are also multiples of π. For example, the second smallest circle has an area of 2π, the next largest has an area of 3π, and so on. Since the areas are increasing at a constant rate, the radiuses must be a series of increasing square roots: $\sqrt{1}$, $\sqrt{2}$, $\sqrt{3}$, $\sqrt{4}$, $\sqrt{5}$, etc.

▶ CIRCULAR CALCULATION (page 17)

One method is to draw lots of tangents. The space that is not drawn in is a reasonable representation of a circle. (A tangent is a line that touches a circle at one point only.)

Another method would be to draw a large number of lines or triangles (as shown in yellow). Although in reality it is a many-sided polygon, the eye interprets it as a circle.

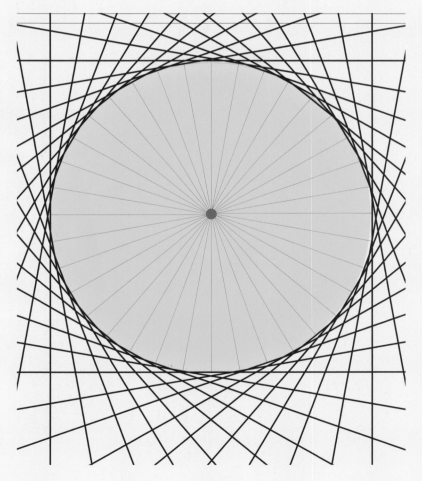

▼ CROSSWAYS (page 18)

The general formula for the number of chords for "n" points is:

$$[n \times (n - 1)]/2.$$

See the solution below to understand why.

(Note the simple number pattern: +1, +2, +3, +4, +5, etc.)

Number of points	1	2	3	4	5	6	7	8	9	10	11	12	13	14	15	16	17	18	19	20
Number of chords	0	1	3	6	10	15	21	28	36	45	55	66	78	91	105	120	136	153	171	190

MYSTIC ROSE—19 POINTS (page 19)

According to the Swiss mathematician Leonhard Euler, it is possible to trace the design in one continuous line because there are no junctions at which an odd number of lines meet.

▶ THE HAMILTONIAN WAY (page 20)

A complete graph on seven points must have Hamiltonian circuits, one of which is shown.

▶ BACK TO SQUARE ONE (page 21)

No Hamiltonian circuit is possible without retracing one of the lines.

▼ THE ICOSIAN GAME (page 21)

There are two distinct solutions to the route-finding problem, as shown.

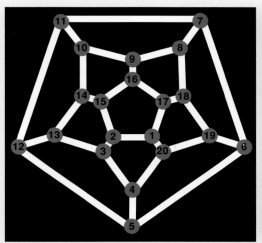

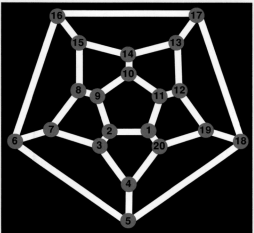

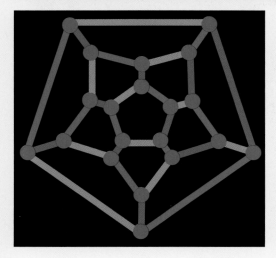

One solution to the coloring problem is given here.

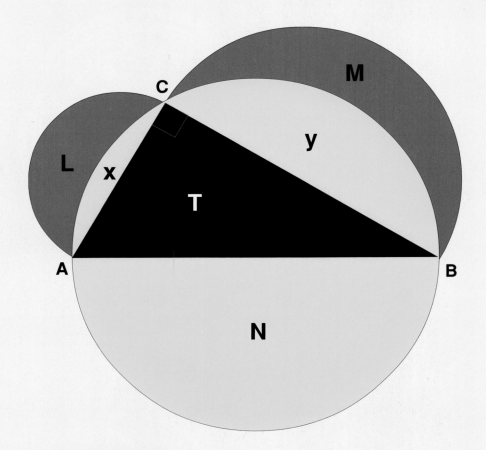

▲ LUNES OF HIPPOCRATES (page 22)

Puzzle 1 By the Pythagorean theorem:
$AB^2 = AC^2 + BC^2$.

By analogy, the areas of the semicircles on the two shorter sides of the triangle will equal the area of the semicircle on the hypotenuse. Hence:
$(L + x) + (M + y) = N$.

However, we note that $N = (T + x + y)$ by symmetry, that is, each half of the circle is identical.

Hence, $(L + x) + (M + y) = T + x + y$. This simplifies to $L + M = T$, which is what we want to know: The area of the triangle T is equal to the sum of the areas of the two lunes, L and M.

Puzzle 2 The sum of the areas of the four crescents is the same as the square. (Divide the square into two right-angled triangles to see why.)

THE CIRCLE INSIDE (page 23)

All the black regions have the same area. It is simple to see that three diagrams are the same except for rearrangement. The third diagram (bottom left) has ¼ of a circle that has twice the radius. Since Area = πr^2, the area is also identical.

▼ LEONARDO'S DISSECTION (page 24)

▼ LEONARDO'S MIRROR (page 25)

Use a mirror to reveal Leonardo's mirror writing:

"Those who are inventors and interpreters between Nature and Man, as compared with the reciters and trumpeters of the works of others, are to be considered simply as an object in front of a mirror in comparison with its image when seen in the mirror, the one being something in itself, the other nothing."

Harsh words!

PRETTY PETALS (page 26)

The four petals are formed by overlapping two semicircles over two semicircles.

Each semicircle has an area of
$\frac{1}{2}\pi r^2 = \frac{1}{2} \times \pi \times 1^2 = \frac{\pi}{2}$.

Hence 4 semicircles $= 4 \times \frac{\pi}{2} = 2\pi$.

The area of 4 semicircles also equals the area of the entire square plus the area of overlap.

Hence, $2\pi = (2 \times 2) + (4\ petals)$.

So 4 petals $= 2\pi - 4$, which is the final answer.

▶ AMPHORA (page 27)

The area of the vase is equivalent to the area of the square:
$(2r)^2 = 4r^2$.

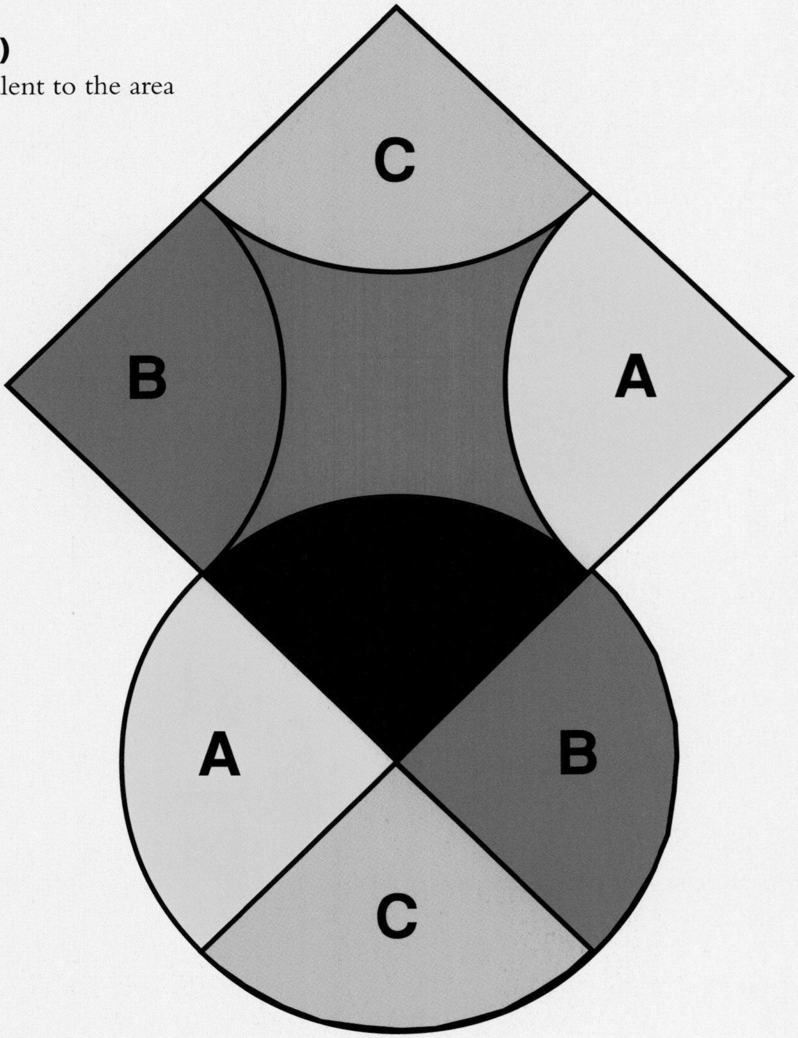

◀ **YIN-YANG (page 27)**
Using this dissection principle shown here, you can divide the circle into as many parts as you wish.

▶ **MAN OF THE EARTH (page 28)**
Eratosthenes knew that sunrays travel in parallel lines and so deduced that the difference in the angles was caused by the curvature of the Earth. Knowing the distance between Alexandria and Syene, which is about 489 miles (787 km), he multiplied this distance by 50 ($360^0/7.2^0$) to determine the circumference of the circle that passes through these two towns and the North and South poles—in other words, the circumference of the Earth.

The full calculation goes: $360/7.2 = X/787$ where X = the circumference of the Earth. Solving the equation, we find X = 24,452 miles (39,350 km) is the circumference of the Earth.

His estimate was remarkably accurate: Today we know that the circumference of the Earth at the Equator is 24,889 miles (40,056 km).

Eratosthenes is also remembered for his prime number sieve, where the prime numbers can be deduced by crossing out every multiple of 2, 3, 5, 7, 11, 13 etc. from a grid of whole numbers.

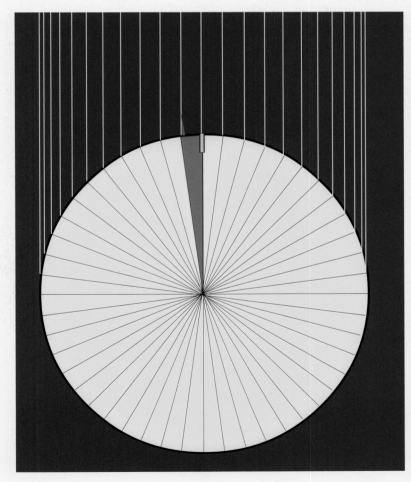

▶ WHAT ON EARTH? (page 29)

Volume of Cylinder − Volume of Sphere = Volume of Cone.

This is the fundamental theorem on which the measurement of a sphere depends. Archimedes considered it one of his greatest achievements.

The ratios of the volumes of a cone, sphere, and cylinder, all of the same height and radius, are in proportion 1:2:3.

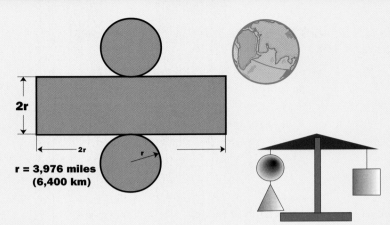

2r

2r

r

r = 3,976 miles (6,400 km)

EARTH (SPHERE) SURFACE AREA—VOLUME

^{A}cylinder $= 2r \times 2\pi r + 2\pi r^2 = 6\pi r^2$

^{A}sphere $= 4\pi r^2 = 4 \times 3.14 \times 3{,}976^2 = 195{,}855{,}715$ mi^2 ($4 \times 3.14 \times 6{,}400^2 = 515{,}000{,}000$ km^2)

^{V}cylinder $= \pi r^2 \times 2r = 2\pi r^3$

^{V}sphere $= \tfrac{4}{3}\pi r^3$

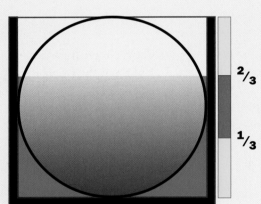

²/₃

¹/₃

◀ HYDROSPHERE (page 30)

To find the answer, consider the quote from Archimedes on page 29.

Volume of sphere $= \tfrac{4}{3}\pi r^3$

$= \tfrac{4}{3}(\pi r^2)r$

$= \tfrac{2}{3} \times (\pi r^2) \times 2r$

$= \tfrac{2}{3} \times$ circular base $\times$ cylinder height

$= \tfrac{2}{3} \times$ cylinder's volume

▶ SPHERE OF INFLUENCE (page 31)

Volume of sphere $= {}^4\!/_3\pi r^3$

$= {}^4\!/_3\pi(\tfrac{D}{2})^3$ where D = diameter of sphere

$= {}^1\!/_6\pi D^3$

$\tfrac{\pi}{6} = \tfrac{3.14}{6}\% =$ approximately 0.52

Hence, water occupies 52% of the cube's volume, as shown.

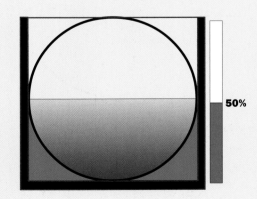

50%

▶ RAMPANT RABBITS (page 33)

Jan	Feb	Mar	Apr	May	Jun	Jul	Aug	Sep	Oct	Nov	Dec
1	2	3	5	8	13	21	34	55	89	144	233

This sequence of numbers above shows the number of rabbit pairs in each month. The total number of pairs at the end of the year is 233.

Can you discover the secret pattern behind this number sequence and continue it further? The answer is that each number is the sum of the two numbers before it.

This sequence, resulting from a manmade recreational puzzle, proved of enormous interest to mathematicians, and even today it plays an important role not only in mathematics, but in science as well.

The "Fibonacci number sequence," as the sequence is today called, has a mysterious relationship with the golden ratio, spiral, rectangle, and triangle, all of which are part of a recurring growth pattern; but the "how" and the "why" still remain a complete mystery.

And all this from a 13th-century puzzle dealing with breeding pairs of rabbits!

SHOWY SUNFLOWER (page 34)

If you count the number of spirals in a sunflower clockwise and then the number of spirals counterclockwise, you will find the two numbers are usually two consecutive Fibonacci numbers: in our case, 21 and 34.

▼ THE MONK AND THE MOUNTAIN (page 35)

This is an interesting problem whose visual representation leads to a fairly obvious solution.

No matter the speed of ascent and descent of the monk during his journeys, or how long he rested, or even if he traveled backward at times, his two paths must intersect at a point somewhere along the route. This is shown in the diagram below, in which the two paths are superimposed.

Here's another way to think about the problem: consider presenting two monks for the two journeys, one walking up and the other walking down at the same time. One starts from the bottom and the other starts from the top at the same time, 7 in the morning. Both arrive at the end of their journey at 7 in the evening. At some point they have to meet regardless of their speed and how often each of them stops. This is the time and place we were asking for.

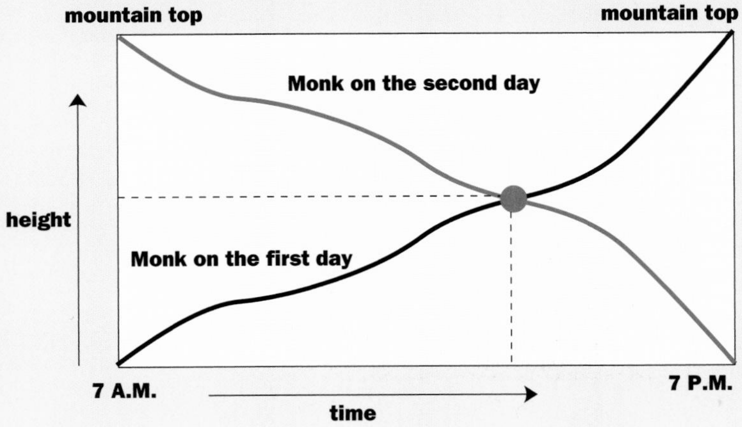

PACKING DISKS OR CIRCLES (page 36)

The best densities are as follows:
1. Square lattice = $\pi/4 = 0.7854$
2. Hexagonal lattice = $\pi/6 \sqrt{3} = 0.9069$

▼ BLACK BALLS (page 38)

Theoretically, after many shakings of the box, the best packing (hexagonal packing) could be reached, which would take about 14% less space in the box than before.

But in practice, when the box is rotated, tapped, or shaken, the spheres assume an infinity of visually compelling configurations, reminiscent of cracks in crystals. The resulting configurations reflect processes in nature.

It can be seen that there are always several regions of regular arrangements, set at angles to each other (grain boundaries), corresponding by analogy to atomic arrangements in crystals, metals, and other materials that occur as a crystal grows.

Key properties of different materials can be determined according to the way in which the atoms are arrayed and the way in which they are joined together. The differences between solids, liquids, and gaseous states are explained by the patterning of their atoms and the relative closeness of their molecules.

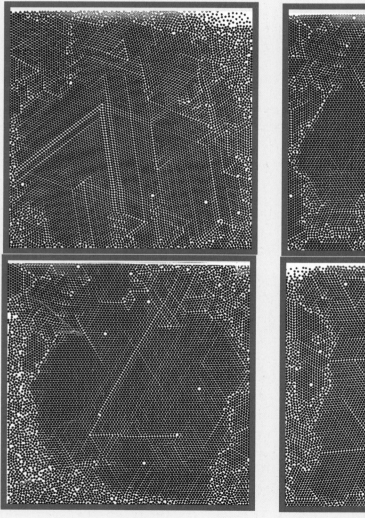
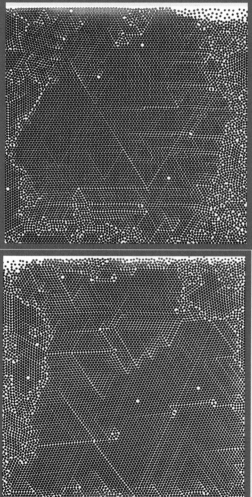

▼ PACK IT IN! (page 39)

Six spheres can be removed as shown, and the box won't rattle.

▶ BOTTLE IT UP (page 40)

You can repack the bottles in hexagonal packing in nine rows as shown, and you will be able to pack two additional bot-tles, making 50 bottles altogether in your crate.

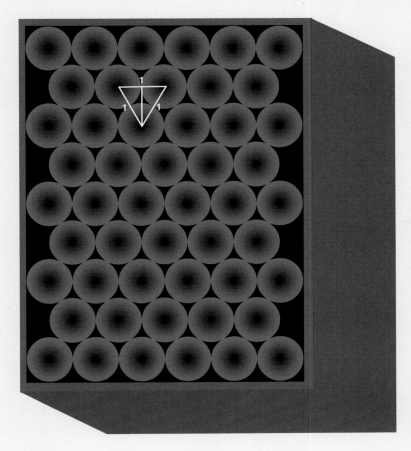

▶ PACKING CIRCLES: TEN IN A SQUARE (page 41)

The radius of the ten identical circles is 0.148204 of the unit squares. It is assumed that no better solution could be found (using larger circles).

This illustrates the best solution so far, found in 1990 by two French mathematicians, Michael Mallard and Charles Payan, and proven by De Croot in 1990.

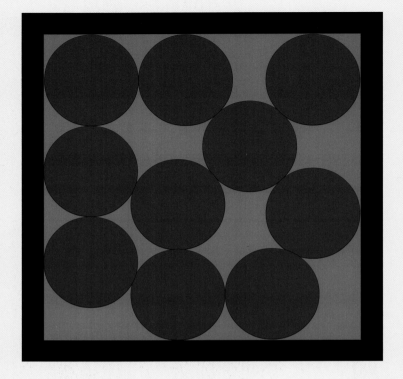

▶ PACKING CIRCLES: 13 IN A SQUARE (page 42)

It may look surprising, but the square of roughly 3.7 diameters is the best solution so far for 13 identical circles, with 12 locked circles, and one circle freely rattling inside, as shown.

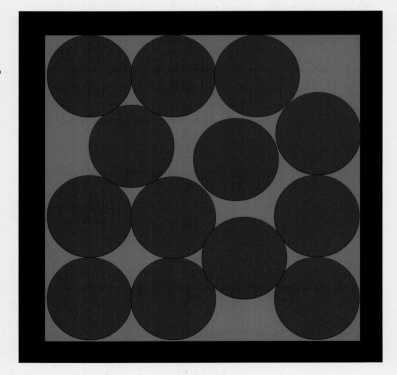

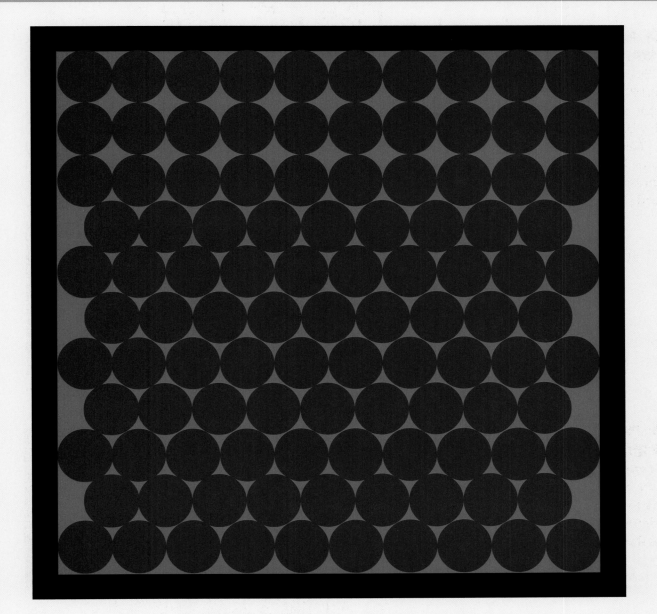

▲ PACKING CIRCLES: 100 IN A SQUARE (page 43)

Yes! By combining the two packings, you can squeeze in another circle, making 106 circles altogether, as shown.

Again, the best solution is not necessarily the most ordered and regular one.

Needless to say, such packing problems are vital for manufacturing, specifically when you want to cut as many circular parts out of a sheet as possible.

▼ MOVE IN DIFFERENT CIRCLES (page 45)

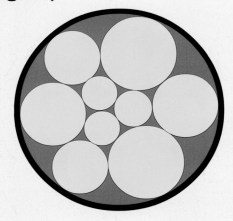

▼ CANNONBALL CLUMP (page 45)

The compressed solid is a rhombic dodecahedron, a beautiful solid with 12 identical faces in the shape of a diamond as shown.

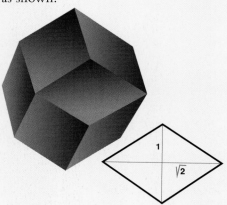

◄ HEXSTEP (page 50)

A 47-move solution is shown. Can you do better?

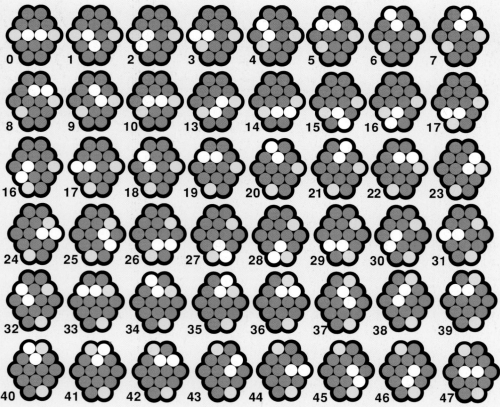

HEXSTEP 2 (page 51)

Puzzle 1 A 33-move solution is shown. Can you do better?

Puzzle 2 A 43-move solution is shown Can you improve on it?

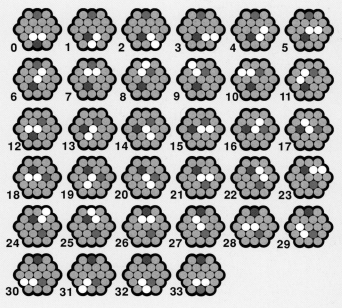

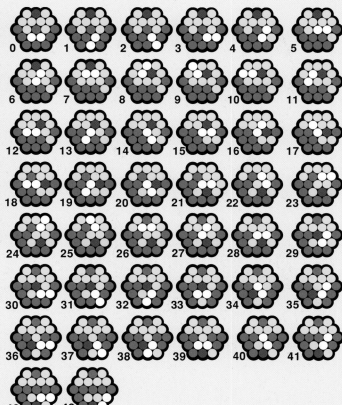

MONGE'S CIRCLES THEOREM (page 52)

No, it is not a coincidence that common tangents from three circles meet on a straight line. It is known as Monge's theorem, for the French mathematician Gaspard Monge (1746–1818). The proof is a good example of the use of the visual imagination and it goes like this:

Imagine three spheres of different sizes resting on a plane. Now imagine three cones, each of which contains two of the spheres.

Plainly their vertices will also rest on the same plane as the spheres.

Now imagine a plane passing through the centers of these three spheres. It will intersect the three spheres and the three cones. But this plane must also contain the three vertices of the cones. Since these vertices all lie on two intersecting planes, they must lie on a straight line.

▶ GOT YOU PEGGED (page 53)

As shown at right, no matter how well you overlap the two plates, six pegs will be left free.

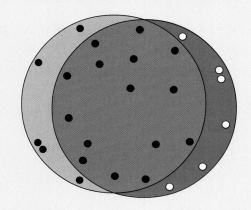

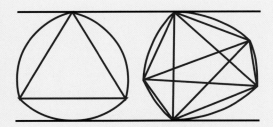

◀ CONVEYANCING (page 54)

It is obvious that the circular roller will not cause us any problems. But what about the others?

We have already seen with the Reuleaux triangle that we can take a triangle and round off the corners so that the shape has a constant width. Therefore, if we were to roll it across a flat table, the height of its highest point will not change. In essence, the shape acts in just the same way as a circle. It turns out that we can do the same thing to the pentagon, as shown. As long as the corners are rounded in the correct manner, a valid curve of equal width is formed. It turns out that any polygon can be treated in this fashion so long as it has an odd number of sides.

Therefore, it follows that the glasses will move smoothly across the conveyor belt.

▶ BUMPY RIDE IN CATENA (page 55)

The Catenarians had bicycles with square wheels, which were just the right size for the bumps in the road (called catenary curves—see page 56).

By now you will have realized that the city of Catena in Sicily is fictional (though there is a city called Catania). And since Catenarians cannot steer on their roads, the only conclusion is that they live in a one-dimensional "Lineland."

But the part about the smooth rides with bicycles with square wheels is true. Stan Wagon built a bicycle with square wheels, shown at right.

CURVES OF CONSTANT WIDTH (page 55)

All except the last turn like a circle.

CATENARY: THE GRAVITY CURVE (page 56)

Tying weights to the chain will increase the tension throughout and thus make the midpoint more rounded. In essence, it turns into something resembling a parabola.

▼ CONIC SECTIONS (page 57)

1. Circle (parallel to base).
2. Ellipse (inclined to the axis at an angle greater than the semivertical of the cone).
3. Parabola (parallel to a generating line of the cone).
4. Hyperbola (inclined to the axis at an angle less than the semi-vertical angle of the cone).

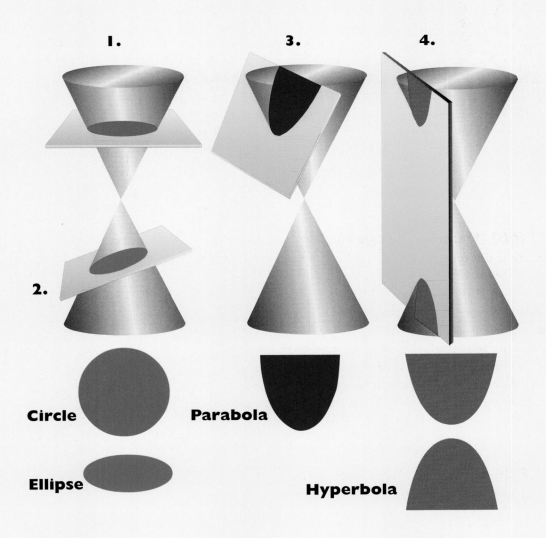

Circle

Parabola

Ellipse

Hyperbola

▶ PITCH UP THE PARABOLA
(page 59)

Parabolas, like circles, always have the same shape, although the shape can be enlarged or diminished. The suspended weights will always form a parabola no matter what the position of the rod.

HOW MANY HEXAGONS?
(page 60)

There is just one shaded hexagon—the red one in the middle. The blue line is a hexagonal spiral line. (The more observant will have spotted that the red border forms a second hexagon, however.)

GOING ROUND IN CIRCLES?
(page 61)

The engraving is actually one single spiral starting from the middle of the nose.

▶ PUZZLE WITH A TWIST
(page 66)

Imagine that you unwrap the cylindrical building and flatten it as shown.

According to the Pythagorean theorem:

$c^2 = a^2 + b^2 = 30^2 + 40^2 = 900 + 1600 = 2{,}500$ units.

Hence, $c = \sqrt{2{,}500} = 50$ units.

Therefore, the staircase length is $4 \times 50 = 200$ units.

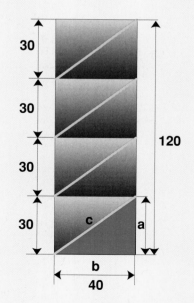

BE FLIPPANT (page 67)

The probability of getting two tails and two heads in order TTHH is the product of the individual probabilties:
$\frac{1}{2} \times \frac{1}{2} \times \frac{1}{2} \times \frac{1}{2} = \frac{1}{16}$.

ELUSIVE ELLIPSE? (page 67)

The ellipse is the curve most often "seen" in objects around us. The man can pick up the glass of water, tilt it, and see the water surface forming a perfect ellipse.

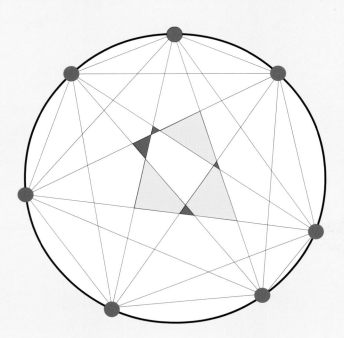

◄ INNER TRIANGLES (page 68)

With seven points to choose from, there are seven possible triangles (as shown). An easy way to see this is to imagine leaving out a different point with each different selection. Leaving out one point leaves six points, which we have seen make one triangle. Seven points can be omitted, thus seven triangles will result.

If we had eight points to choose from $(8 \times 7)/2 = 28$ triangles would be possible.

►PHOTO-GRAPHIC (page 69)

You can draw 12 edges as shown.

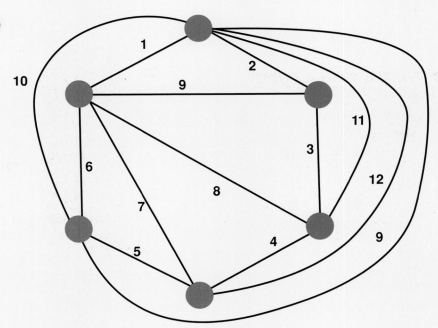

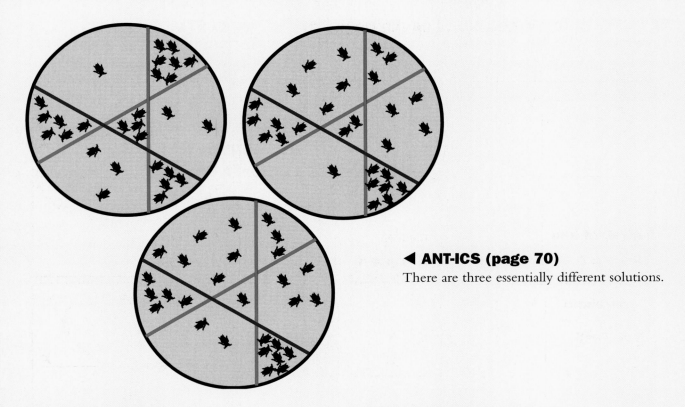

◀ ANT-ICS (page 70)

There are three essentially different solutions.

**▼ BUTTERFLY COLLECTION
(page 71)**

▼ WEB WEAVERS (page 71)

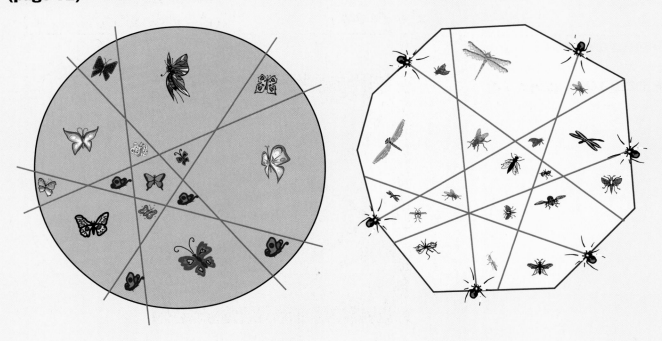

▼ CURIOUS CURVES (page 72)

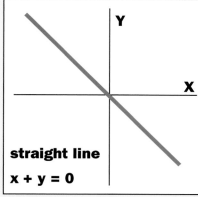

straight line
x + y = 0

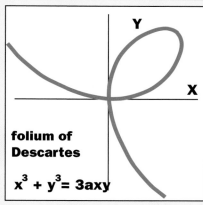

circle
$x^2 + y^2 = 4$

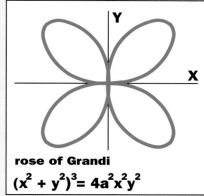

ellipse
$x^2 + 4y^2 = 4$

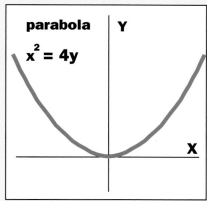

parabola
$x^2 = 4y$

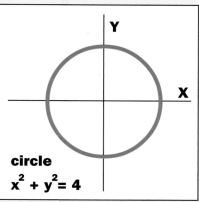

folium of Descartes
$x^3 + y^3 = 3axy$

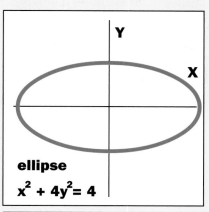

rose of Grandi
$(x^2 + y^2)^3 = 4a^2x^2y^2$

▶ GRID LOCK (page 73)

The image is of a rabbit in a bow tie.

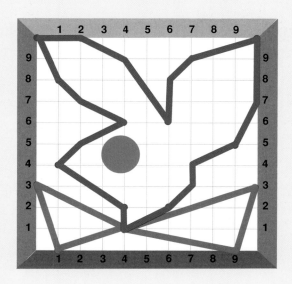

▼ MAGIC COLOR SQUARES QUARTET (page 74)

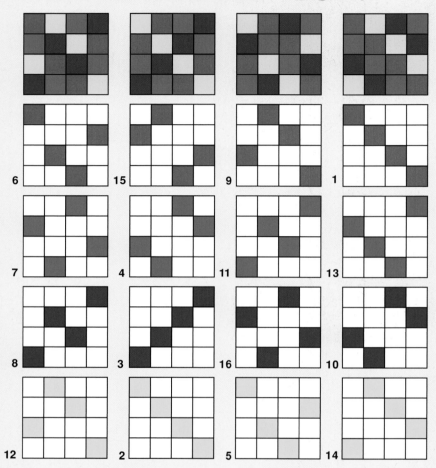

▼ RED–GREEN–BLUE (page 75)

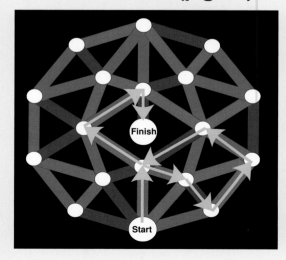

▼ GREEN–RED–BLUE (page 75)

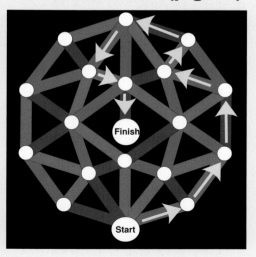

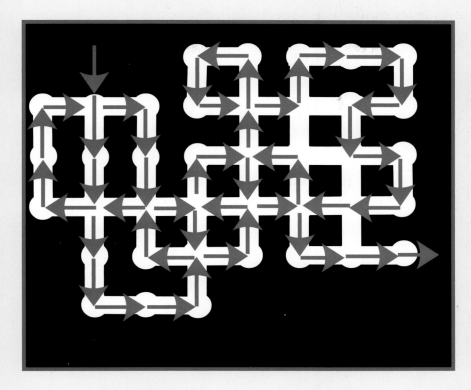

◀ CAVE IN, THEN OUT
(page 78)

RECONNAISSANCE
(page 79)

You can't do it!

After passing an odd number of intersections, you will be inside the boundary.

To return to the starting point, you must have passed an even number of intersections. Thus the number can't be 11 or any other odd number.

A DOTTY ACTRESS
(page 80)

There are 772 dots forming Marilyn.

▶ ALPHABET SOUP
(page 81)

Group 1: The red letters are capital letters of the alphabet with vertical symmetry. The blue letters are those with horizontal symmetry.

If a letter is bilaterally symmetrical you can find at least one line (or maybe more) that will bisect it into two mirror-image halves.

The mirror-image of the capital letter "A" can cover the original letter exactly in the plane. However, there are some letters that are not symmetric and consequently they cannot be covered in the plane by their mirror-images.

Group 2: The blue letters are capital letters of the alphabet with both horizontal and vertical symmetry axes. The red letters are asymmetrical capital letters with no closed areas.

Group 3: The red letters are asymmetrical and contain closed areas, while the blue letters have a two-fold rotational symmetry.

Some letters may have no bilateral symmetry (no line that will dissect them into two mirror-halves), but still possess rotational symmetry.

Group 1

A C T E Y K D U B W

Group 2

H F J O I G L X I

Group 3

N R S P Q Z

▶ ANY WHICH WAY
(page 81)

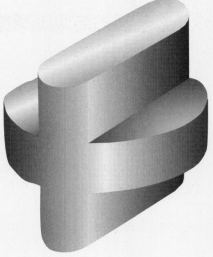

SKELETON CUBES (page 82)

We can choose a color for each side in turn, that is, $12 \times 11 \times 10 \times ... \times 3 \times 2 \times 1 = 479,001,600$.

However, the cube can lie on six sides and each side can have four orientations, so the true number of different cubes is $(479,001,600) \div (6 \times 4) = 19,958,400$.

PANTOGRAPH (page 83)

The red point will trace a line enlarged 2.5 times.

This is the principle of an ancient instrument called the pantograph, based on Euclidean geometry, which has been around for many centuries or more. Leonardo da Vinci used it to duplicate his drawings.

Initially it was used for copying documents to a desired scale—reduced or enlarged. Basically it is a plane linkage based on the geometry of a parallelogram (it is in fact a parallelogram with two extended links).

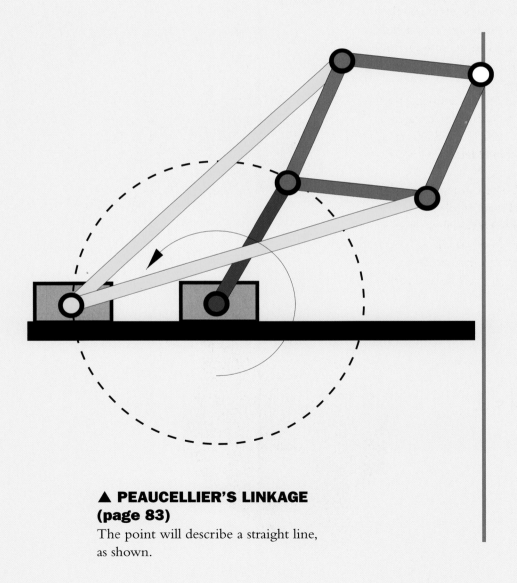

▲ PEAUCELLIER'S LINKAGE (page 83)

The point will describe a straight line, as shown.

FLATTERY WILL GET YOU NOWHERE
(page 89)

Initially, the Flatlanders will not be aware of the cube's approach because they cannot see anything outside of their universe. What happens thereafter somewhat depends on how the cube approaches Flatland. In all cases, Flatlanders will see only a red two-dimensional shape appear in their world.

Case 1: If, as illustrated, the cube leads with its sharp point, the Flatlanders will see: a dot, then a growing triangle, then a hexagon that becomes gradually more regular, and then the sequence is reversed until the cube has passed right through.

Case 2: If the cube makes first contact along one edge, they will see a line that grows into a series of increasingly wide rectangles, and then the sequence is reversed.

Case 3: If the cube passes through with its lowest face parallel with Flatland, the Flatlanders will see a solid square (without variation in size) suddenly appear for a few moments before disappearing once more.

Of course, if the cube is spinning as it passes through, the situation becomes even more complex!

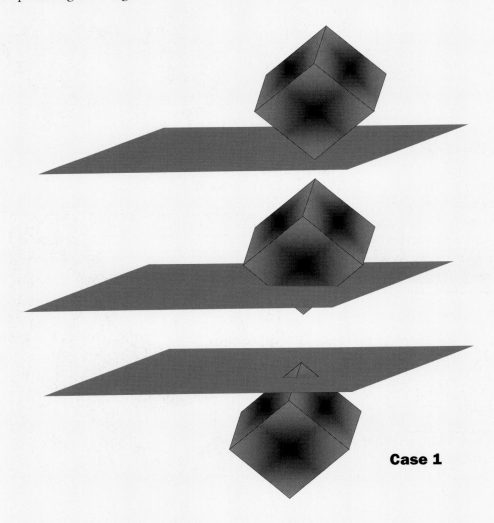

Case 1

▶ BANK RAID
(page 90)

This is how a Flatlander can open the vault and remove anything within it.

On the other hand, a three-dimensional person could easily remove the gold without breaking or opening the sealed vault. By analogy, a four-dimensional person could achieve this feat in our world and empty all the gold stored in banks anywhere.

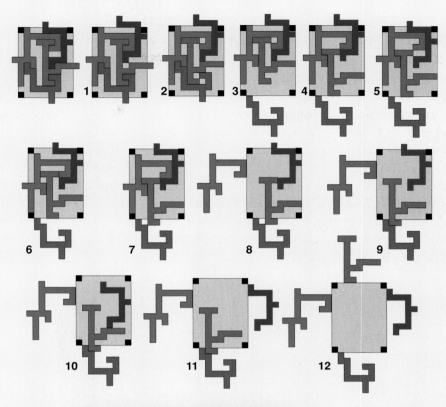

▼ ON THE RIGHT TRACKS
(page 91)

The best solution is shown, and creates 32 intersections. It must be the best solution because if you were to take one of the seven horizontal tracks and lay it diagonally instead, you would lose four intersections but create eight new ones.

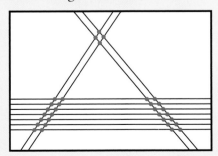

▼ ARE YOU COORDINATED?
(page 92)

The picture shows a seal.

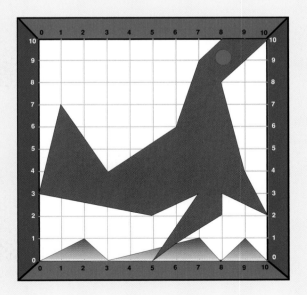

▶ CAT'S CRADLE
(page 93)

Overlapping the string we can create a straight line of one-unit length, a 90-degree right angle, as well as a circle, as shown. Can you find more?

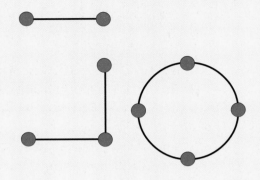

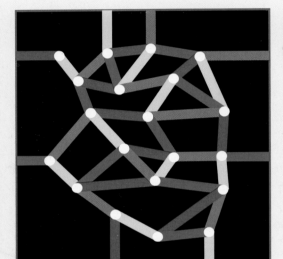

◀ THE PATH TO THE FINISH
(page 95)

▶ MOUTH TO MOUTH
(page 96)

This rearrangement (with a face vanished) makes 12 faces as required—five smiling and seven sad.